BIODEGRADATION OF HAZARDOUS AND SPECIAL PRODUCTS

Edited by **Rolando Chamy**
and **Francisca Rosenkranz**

Biodegradation of Hazardous and Special Products
http://dx.doi.org/10.5772/52778
Edited by Rolando Chamy and Francisca Rosenkranz

Contributors

Md.Azizul Moqsud, Lebin Yin, Nina Djapic, Yasko Kodama, Yu-Huei Peng, Yang-Hsin Shih, Anita Rani Santal, N. P. Singh, Hugo Ramirez Saad, Ildefonso Díaz-Ramírez, Erika Escalante-Espinosa, Randy Adams Schroeder, Reyna Fócil Monterrubio, Urszula Guzik, Katarzyna Hupert-Kocurek, Danuta Wojcieszyńska

Published by InTech

Janeza Trdine 9, 51000 Rijeka, Croatia

Publishing Process Manager Iva Lipovic
Technical Editor InTech DTP team
Cover InTech Design team

Additional hard copies can be obtained from orders@intechopen.com

Biodegradation of Hazardous and Special Products, Edited by Rolando Chamy
and Francisca Rosenkranz
p. cm.
ISBN 978-953-51-1155-9

For EU product safety concerns:
IN TECH d.o.o., Prolaz Marije Krucifikse Kozulić 3, 51000 Rijeka, Croatia,
info@intechopen.com or visit our website at intechopen.com.

Meet the editors

Rolando Chamy is a Biochemical Engineer and Ph.D. in Chemical Engineering from the University of Santiago de Compostela, Spain. He holds a professor chair at the Pontificia Universidad Católica de Valparaíso, and is the Director of the Biotechnology Center of Curauma, a center of applied research, which belongs to the same university. Currently, he also leads the research line in Renewable Bioresource at Fraunhofer Chile Research. His main lines of research are environmental biotechnology, renewable energy, climate change, and innovation management.
Additionally, he has developed an extensive body of work in the field of applied research, with more than 100 as project director, principal investigator or co-investigator, supported with more than 100 ISI publications and 19 contributions to book chapters.

Francisca Rosenkranz, Biochemical Civil Engineer (Pontificia Universidad Católica de Valparaíso) and PhD in Biotechnology (Pontificia Universidad Católica de Valparaíso and Universidad Técnica Federico Santa María) with the co-supervision of the University of Santiago de Compostela, Spain. Her specialization areas include biogas production, anaerobic treatment of hard biodegradable compounds and studies of microbial communities in anaerobic systems. Her actual positions are at Fraunhofer Chile Research, as research project coordinator in the bioenergy area, and at Biotechnology Center of Curauma, as Chief of Operations.

Contents

Preface

This book contains a collection of different research activities that include the biodegradation compounds with contaminant characteristics and special products of different interests as an added value product or that allows following up various biological processes.

The chapters consider the degradation of contaminant compounds generated by industrial activities, i.e., oil industry by-product compounds and halogen compounds or compound generated by natural phenomena such as tsunamis, which require interventions to recover damaged soils.

In addition, the book contains chapters that involve special product degradation processes such as chlorophyll, which corresponds to a biological process indicator as photosynthesis.

Rolando Chamy and Francisca Rosenkranz
Biotechnology Center of Curauma
Pontificia Universidad Católica de > Valparaíso, Chilr

School of Biochemical Engineering
Pontificia Universidad Católica de > Valparaíso, Chile

Chapter 1

Isolation and Characterization of Cypermethrin Degrading Bacteria Screened from Contaminated Soil

L. B. Yin, L. Z. Zhao, Y. Liu, D. Y. Zhang,
S. B. Zhang and K. Xiao

Additional information is available at the end of the chapter

http://dx.doi.org/10.5772/56339

1. Introduction

A current environmental concern is the contamination of aquatic ecosystem due to pesticide discharges from manufacturing plant, agricultural runoff, leaching, accidental spills and other sources [1, 2]. Synthetic pyrethroid insecticides were introduced into widespread use for the control of insect pests and disease vectors more than three decades ago. In addition to their value in controlling agricultural pests, pyrethroids are at the forefront of efforts to combat malaria and other mosquito-borne diseases [3] and are also common ingredients of household insecticide and companion animal ectoparasite control products [4]. Cypermethrin is a type of synthetic pyrethroids (SPs), a class of pesticides widely used for insect control in both agricultural and urban settings around the world [5].

The use of SPs in China has increased sharply since many organophosphate products, such as methamidophos and parathion, are being phased out for agricultural use. With such extensive application, many adverse effects, such as pest resistance, residues in foods, and environmental contamination are public safety concerns [6, 7]. Although SPs are widely considered safe for humans, numerous studies have shown that exposure to very high concentrations of SPs might cause human health problems [8]. Such effects include bioaccumulation toxicity; immune suppression, endocrine disruption; modify electrical activity in various parts of the nervous system, neurotoxicity, lymph node and splenic damage, and carcinogenesis [9-11]. In addition, bees, fish, crabs, tadpoles, arthropods, and other non-target organisms are extremely sensitive to the toxic effects of SPs [12-15].

Cypermethrin is more effective against pests including moth pests of cotton, fruits and vegetable crops. Extensive and improper use of this kind chemicals leads to greater health risk to plants, animals and human population which had been reviewed time to time by several researchers [16]. One of the major problems asides from toxicity and carcinogenicity of pesticides is their long persistence in nature that amplifies the toxicity and health risk problems in the area of contamination [17].

Therefore, it is necessary to develop a rapid and efficient disposal process to eliminate or minimize the concentrations of SPs in the environment. A variety of physical and chemical methods are available to treat the soils contaminated with hazardous materials but many of these physical and chemical treatments do not actually destroy the hazardous compounds but are bound in a modified matrix or transferred from one phase to another [18, 19], hence biological transforming is essential. The biological treatment of chemically contaminated soil involves the transformation of complex or simple chemical compounds into non-hazardous forms [20]. For biodegradation, ideally the target pesticide will be able to serve as the sole carbon source and energy for microorganisms, including the synthesis of appropriate enzymes if need able. The specificity of enzymes active against xenobiotic compounds differs from one microorganism to another.

In the light of this fact, biodegradation, especially microbial degrading, has proven to be a suitable method for insecticide elimination. Previous studies indicated that microbes play important roles in degrading and detoxifying SPs residues in the environment. Thus far, many reports have described the biodegradation of cypermethrin by various bacteria, including *Ochrobactrum lupini*, *Pseudomonas* aeruginosa, *Streptomyces aureus*, and *Serratia* spp.[21-23], but there is few research describing biodegradation of pesticides by *Rhodobacter sphaeroides*. Among the different genera of pesticide-degrading bacteria, the photosynthetic (PSB) genus *Rhodobacter* has a special status in the ecosystem, since its metabolic functions are extraordinarily versatile, including degradation of various organic compounds, nitrogen fixation, hydrogen production [24], as a biofertilizer for promoting plant growth and increasing grain yield [25], and 5-aminolevulinic acid production which has multiple functions including a relatively strong herbicidal effect in clover [26]; therefore, microbes belonging to this genus are ideal choice for degrading pesticide residues.

The research aim was to identify the potential microbial strain able to utilize cypermethrin from the contaminated soil. In this study, the pesticide degrading potential of a bacterial culture is examined with the hope of isolation and characterization of cypermethrin degrading potentials in the contaminated soil. In addition, the optimum dose and the suitable conditions for cypermethrin degradation using laboratory scale were also evaluated. The results of the present study suggest that the use of potential microorganisms in the treatment system can successfully overcome many of the disadvantages associated with the conventional method used for the degradation of inhibitory compound.

2. Materials and methods

2.1. Chemicals and media

Standard analytical grade sample of 100 μg/mL cypermethrin (99.8% purity) was purchased from the Agro-Environmental Protection Institute, Ministry of Agriculture (Tianjin, China). Acetonitrile, methanol and hexane were of chromatographic grade while other chemicals were of analytical grade. Cypermethrin dissolved in acetone solution was added to desirable concentration in medium as the sole carbon source. Mineral Salts Medium (MSM) (g/L): 1.0 NH_4NO_3, 1.0 NaCl, 1.5 K_2HPO_4, 0.5 KH_2PO_4, 0.2 $MgSO_4$ $7H_2O$, pH 7.0. For solid plate, 1.5% (w/v) agar was added. Medium were sterilized by autoclaving at 121℃ for 30 min before use.

2.2. Enrichment, isolation and screening of bacterial strains

An activated sludge sample was collected from the wastewater treatment pool of a pesticide plant located in Changsha (Hunan, China), which had produced cypermetrin over 5 years. Wastewater sludge enrichment was performed by placing 10 g activated sludge in a 250 mL-Erlenmeyer flask containing 100 mL sterilized MSM media with an initial cypermethrin concentration at 20 mg/L, and incubated in a light incubator (PRX-450D, China) at 37℃ and 7500 lux; the flasks were shaken 3–5 times per day. After 10 days or so, the medium turned red-brown, a 5 mL aliquot of the culture was inoculated into 100 mL of fresh MSM medium containing 50 mg/L cypermetrin, and the new mixture was incubated for another 10 days under the same conditions. The medium was gradually acclimated to increasing concentrations of cypermetrin ranging from 50 to 200 mg/L at intervals of a week. After about 10 transfers, a mixed microbial population was diluted in series, and then streaked on MSM agar medium plate containing 100 mg/L cypermethrin. The dilution series was repeated at least 5 times, until single colony was achieved. The abilities of isolates to degrade cypermethrin were determined by gas chromatography (GC) according to Yin et al and Chen et al [23, 27]. The relatively higher degradation ability colonies were selected for further degrading studies. These organisms were stored long-term on porous beads in a cryopreservative fluid at -20℃ and short-term on agar plates at 4℃

2.3. Characterization and identification of the cypermethrin degrading isolates

A cypermethrin degrading isolate designated as S_{10-1} showed the highest degradation rate was selected for further study. The purified S_{10-1} was identified on the basis of its morphological characteristics and results of biochemical tests and 16S rRNA gene sequence analysis. The isolate S_{10-1} was grown on MSM agar plates containing 50 mg/L cypermethrin at 37℃ and 7500 lux for 7 days, its cell morphology, method of reproduction, and the structure of its inner photosynthesis membrane and flagella were observed by transmission electron microscope (JEM-6360, JEOL) and/or scanning electron microscope (JSM-6360LV, JEOL).

The isolate S_{10-1} was further confirmed by 16S rRNA gene sequence. The DNA was extracted and purified using the Qiagen genomic DNA buffer set. PCR amplification was performed as described by Mirnejad et al [28]. The 16S rRNA sequencing was performed by Beijing Liuhe

Huada Genomic Company (Beijing, China). The sequences with the highest 16S rDNA partial sequence similarity were selected and compared by CLUSTAL W. Phylogenetic and molecular evolutionary analyses were conducted by MEGA 4.0 software with the Kimura 2-paremeter model and the neighbor joining algorithm [29]. Confidence estimates of branching order were determined by bootstrap resampling analysis with 1000 replicates.

2.4. Inoculum preparation

Unless otherwise stated, the inoculants for this experiment were bacteria cultured in a 130 mL serum bottle containing 120 mL of PSB medium in a light incubator at 35℃ and 7500 lux. At the exponential phase (about 2–3 days), the cell pellets were harvested via centrifugation (5000×g, 10 min), washed 3 times with 50 mL of KH_2PO_4-K_2HPO_4 (0.15 mol/L, pH 7.0), and then suspended in the same phosphate buffer as the inoculants. In order to avoid the effects of hydrolysis and photolysis, each treatment was set in triplicate with non-inoculated samples as control under the same conditions and analyzed in the same manner. Samples for residual pesticide concentration analysis were collected from the cultures at regular intervals.

2.5. Optimal conditions for degrading cypermethrin by S_{10-1}

To determine the optimal conditions for degrading cypermethrin by S_{10-1}, single-factor test was designed in this study under different conditions. To confirm the effects of temperature on degradation, the media were placed in illuminating incubators at 10, 20, 25, 30, 35, and 40℃, respectively. To determine the effect of cypermethrin concentration on degradation, MSM media were added with cypermethrin ranging in concentration from 100 mg/L to 800 mg/L. The media were prepared at pH values from 4.0 to 11.0 buffers for the measurement of the effects of pH on degradation. All experiments were conducted in triplicate. The non-inoculated controls throughout the studied were implemented at the same condition in order to exclude the abiotic degradation affection.

2.6. Extraction of cypermethrin for residue analysis

The extraction and quantification of cypermethrin residue in the media was modified slightly from method described in Yin et al [27] and Liu et al [30]. At different time intervals, triplicate populations were sampled for cypermethrin concentration analysis. Cypermethrin was extracted three times from the media with 100 mL of hexane. The hexane extracts from the same samples were combined, dried with anhydrous sodium sulfate, and concentrated by exposure to nitrogen gas to near dryness on a rotary evaporator at room temperature, and then dissolved in 5 mL of hexane for GC detection. Before detection the residues were purified using hexane pre-poured Florisil® columns (Agilent SAMPLIQ Florisil®, USA) and 0.22 μm membranes (Millipore, USA), and were then recovered in 5 mL of hexane; finally, the residues were analyzed by performing GC. Preliminary experiments showed that the recovery of cypermethrin in the above extraction and analysis procedures was >90%.

Residue analyses of cypermethrin degradation were performed using an Agilent 6890N GC system (Agilent Technologies, USA) equipped with an electron capture detector (μ-ECD); an

HP-5 5% phenyl methyl siloxane capillary column (30 m × 320 μm × 0.25 μm; Agilent Technologies, USA) was used for separation, with helium as the carrier gas (flow rate, 1 mL/min). Other GC parameters included an inlet temperature of 250℃ and a detector temperature of 300℃; initially, the oven temperature was 150℃ for 2.0 min, was ramped to 280℃ at 15℃/min, and then maintained at 280℃ for 5.0 min. The injection volume was 1.0 μL. Samples were introduced in split-less mode. Concentrations were determined by analyzing peak area with an authentic cypermethrin standard.

2.7. Detection of cypermethrin metabolites

Metabolites were isolated from the culture filtrates of the organism grown in cypermethrin (100 mg/L) by extraction with acetonitrile, before and after acidification to pH 2 with 2 M HCl, and the residue obtained was dissolved in hexane [22]. The metabolites were identified and analyzed using the GC/MS system (Agilent 7890A/5975, Agilent Technologies, USA) equipped with electron ionization (EI). EI (70 eV) was performed with a trap current of 100 mA and a source temperature of 200℃. Full scan spectra were acquired at m/z 45–500 at 2 sec per scan. The metabolites were confirmed by standard MS, data collection and processing were performed using Agilent MSD ChemStation software containing the Agilent chemical library.

3. Results and discussion

3.1. Isolation and characterization of cypermethrin degrading bacterium

After repeated enrichment and purification processes, we obtained approximately 20 strains of organisms with different colony morphologies from the activated sludge samples. But the degradation experiments showed the isolate S_{10-1} possessed the relatively higher degradation, capacity of degrading cypermethrin (100 mg/L) by 90.4% after incubating 7 days at pH 7.0 and temperature 35℃ (Fig. 3a). And S_{10-1} utilized cypermethrin as its sole carbon and energy source in MSM. Thus strain S_{10-1} was selected for further detail investigation.

S_{10-1} is a gram-negative, anaerobic bacterium. The morphology of the S_{10-1} colonies, cultured for 10 days on MSM agar plate, were reddish-brown, smooth, circular, wet, nontransparent, glistening, and with entire margins (Figure 1a). The physiological and biochemical characteristics of S_{10-1} are shown in Table 1. SEM observations showed that the cells are ovoid to rod shaped (Figure 1b), sometimes even longer, measuring about 0.5–0.9 μm in width and 1.2–2.0 μm in length, and are motile by means of polar flagella (Figure 1c). Internal photosynthetic membranes appear as lamellae underlying and parallel to the cytoplasmic membrane (Figure 1d). The culture suspension was reddish-brown in color. In vivo absorption maxima of intact cells (Figure 1e) were recorded at 378, 455, 480, 510, 592, 806, and 865 nm, indicating the presence of bacteriochlorophyll a and carotenoids of the spheroidene series [31]. These morphological and biochemical properties are identical to the genus *Rhodobacter* [31].

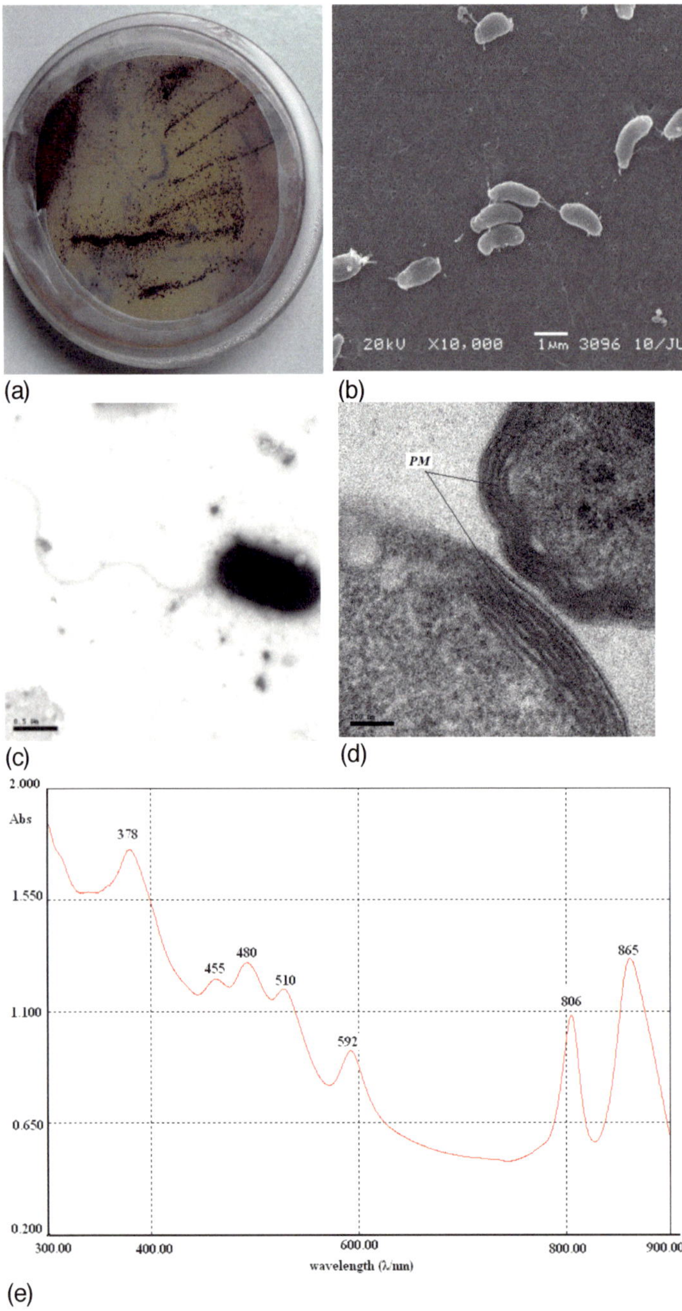

Figure 1. The characterization of strain $S_{10\text{-}1}$. (a) The morphology of the $S_{10\text{-}1}$ colonies, cultured for 10 days on MSM agar plate; (b) Scanning electron micrograph of strain $S_{10\text{-}1}$ (10,000×); (c) Electron micrograph of negatively stained $S_{10\text{-}1}$ cells showing polar flagella (40,000×); (d) Transmission electron micrograph of $S_{10\text{-}1}$: a cross-section showing the photosynthetic membrane (*PM*) lying parallel to the cytoplasmic membrane (200,000×); (e) Absorption spectra of living $S_{10\text{-}1}$ cells.

Items	Results	Items	Results	Items	Results
Gram stain	-[a]	3% NaCl	-	Aerobic dark growth	+
Motility	+[b]	M. R reaction	-	Succinate utilization	+
Hydrogen sulfide	+	Citrate utilization	+	Mannitol utilization	-
V–P reaction	-	Acid from carbohydrates	-	Glycerol utilization	+
Gelatin liquefaction	+	Indole production	-	Pyruvate utilization	+
Catalase	+	Urease	-	Benzoate utilization	-
Oxidase	+	Pigment production	+	Ammonia utilization	+
Strach hydrolysis	+	Nitrate reduction	+	Tartrate	-

Note: [a] Negative/Substrate not utilized; [b] Positive/Substrate utilized.

Abbreviation: VP-Vogues Proskauer; MR-Methyl Red.

Table 1. Physiological and biochemical characteristics of $S_{10\text{-}1}$

3.2. Phylogenetic analysis and identification of $S_{10\text{-}1}$

A 1380-bp 16S rRNA fragment was amplified from the genomic DNA of $S_{10\text{-}1}$ and sequenced (Genebank Accession NO. HM193898). Phylogenetic analysis of 16S rRNA revealed that $S_{10\text{-}1}$ belonged to the genus *Rhodobacter sphaeroides* (Figure 2). $S_{10\text{-}1}$ was temporarily identified as *R. sphaeroides* according to its morphology, colony and cultural properties, physiological and biochemical characteristics, absorption spectra (living cells), internal photosynthetic membrane, and phylogenetic analysis.

Microbial belong to the genus *Rhodobacter*, which are known to play a major role in the treatment of organic wastewater, since they can utilize a broad range of organic compounds as carbon and energy sources; moreover, they are ubiquitous in fresh water, soil, wastewater, and activated sludge. Thus they have been selected for the treatment of many types of wastes [32-34], while *R. sphaeroides* appears to be a new bacterium that may participate in efficient degradation of cypermethrin. To our knowledge, there is not any information concerning the ability of *R. sphaeroides* to degrade cypermethrin and other SPs. However, reports showed that *R. sphaeroides* could effectively degrade pesticides including 2,4-d, quinalphos, monocrotophos, captan and carbendazim [35].

3.3. Effect of temperature on cypermethrin degradation in MSM

Cypermethrin was degraded by $S_{10\text{-}1}$ during incubation temperatures ranging from 10℃ to 40℃. The cypermethrin residues were detected after 7 days' treatment. In cultures incubated at 10℃ and 20℃, the results show that the degradation rate were 15.3% and 23.8%. However, in cultures incubated at higher temperature, i.e. 25℃, 30℃ and 35℃, the degradation rate reached 70.4%, 87.4% and 90.4% within 7 days, respectively, but the degradation rate was only

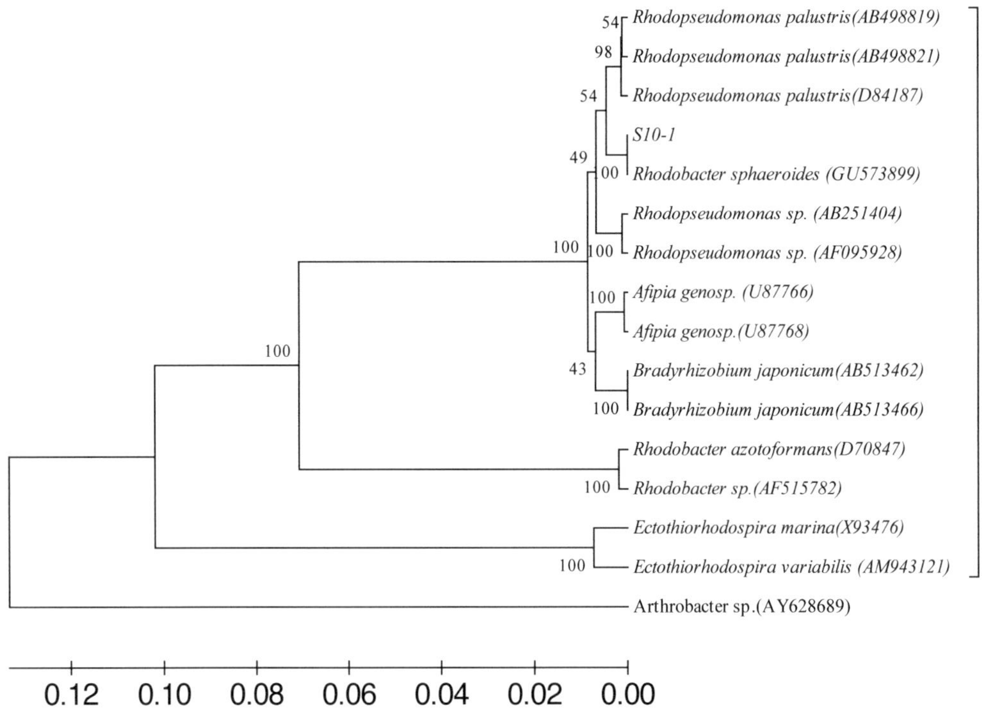

Figure 2. Phylogenetic tree constructed by the neighbor-joining method based on 16S rDNA sequences of S_{10-1} and related strains. Bootstrap values are given at branching points. The sequence of *Arthrobacter* spp. (AY628689) was selected as an out group. The tree was constructed using the neighbor-joining method. Bootstrap values at nodes were calculated using 1,000 replicates (only values >70% are indicated). The GeneBank accession numbers for 16S rRNA gene sequences are shown in parentheses.

61% when incubated at temperature 40℃ for 7 days. The best temperature for degradation was 35℃ (Figure 3a). Similar results were reported by Lin et al [36] who reported temperature significantly influenced cypermethrin degradation by *Streptomyces* sp. strain HU-S-01. Our results also reveal that cypermethrin degradation occurred at 30–35℃ indicating strain S_{10-1} preferred relatively high temperature condition. These results were consistent with previous findings of Chen et al [21]. It is possible that some key enzyme(s) responsible for cypermethrin degradation have their optimum enzymatic activity over such range of temperature. In non-inoculated controls at different temperatures, abiotic degradation was negligible throughout the studies.

3.4. Effect of initial concentration on cypermethrin degradation in MSM

Cypermethrin degradation at different initial concentrations by strain S_{10-1} was investigated. The cypermethrin degradation rates were found to be 90.4%, 60.3%, 38.4%, 32.3%, and 28.7% at concentrations of 100, 200, 400, 600, and 800 mg/L, respectively (Figure 3b). At low cypermethrin concentration ranging from 100 to 200 mg/L, the degradation rate reached above 60%

within 7 days. However at high concentration (400 to 800 mg/L), only about 30% was degraded within 7 days. It might be because of the fact that microbial degradation starts slowly and requires an acclimation period before rapid degradation occurs at high concentration. Similar results were reported by Lin et al [36] who reported that initial concentration of carfofuran was significantly efficiently degraded by *Pichia anomala* strain HQ-C-01 in contaminated soils. In non-inoculated controls at different initial concentrations, abiotic degradation was negligible throughout the studies.

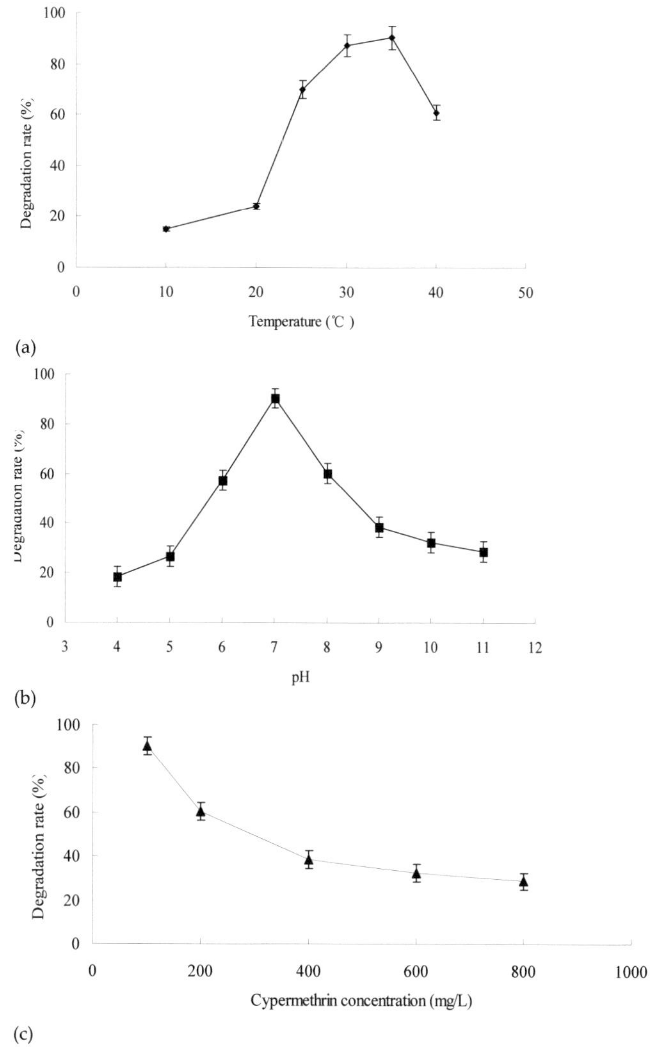

Figure 3. Optimal conditions for degrading cypermethrin by $S_{10\text{-}1}$. (a) Effect of temperature on the degradation of cypermethrin by $S_{10\text{-}1}$; (b) Effect of the initial cypermethrin concentration on the degradation by $S_{10\text{-}1}$; (c) (d) Effect of pH on the degradation of cypermethrin by $S_{10\text{-}1}$. Error bars represent standard deviation (SD) from the mean. Error bars smaller than symbols are not depicted.

3.5. Effect of pH on cypermethrin degradation in MSM

The pH is also an important factor, which significantly effects the degrading ability of bacteria capable of degrading toxicities [37, 38]. To determine the effect of pH on degradation, MSM medium prepared with different pH buffers, fortified with 100 mg/L cypermethrin, and incubated at 35℃ and 7500 lux. Eight different pH (4.0, 5.0, 6.0, 7.0, 8.0, 9.0, 10.0, 11.0) were tested in the optimization experiment. The result showed that the degradation rate were 18.5%, 26.7%, 57.5%, 90.4%, 60.3%, 38.4%, 32.3%, 28.7%, respectively (Figure 3c). The optimal initial pH value for degradation was between 6.0 and 8.0. Results revealed that $S_{10\text{-}1}$ was able to degrade cypermethrin over a wide range of pH. Similar results were reported by Zhang et al [36] who reported that initial pHs were significantly efficiently degraded by two *Serratia* spp., and rapid degradation of cypermethrin at high pH while it was relatively low at acidic pH. In non-inoculated controls at different pH conditions, abiotic degradation was negligible throughout the studies.

3.6. Identification of cypermethrin degradation metabolites

The degradation metabolites of cypermethrin by strain $S_{10\text{-}1}$ were extracted and identified by GC/MS using Agilent MSD ChemStation software containing the Agilent chemical library. GC/MS analysis of the metabolites showed the presence of 4 products. These compounds corresponded with cyclopropanemethanol (Figure 4a), 5-methoxy-2-nitrobenzoic acid (Figure 4b), 3,5-dimethoxybenzamide (Figure 4c), and 5-aminoisophthalic acid (Figure 4d). The retention times of these compounds were 13.609, 14.874, 16.980, and 17.323 min, respectively.

Previous studies had reported about the biodegradation pathway of SPs [21, 39, 40]. In the molecular structure of SPs there is an ester bond which is not as firm as other chemical bonds. Literature indicated that the first step in the microbial degradation and detoxification of SPs is the hydrolysis of its carboxyl ester linkage [23, 36, 41]. However, the chemical bond broken of cypermethrin metabolites are not detected as that described in a previous study. It is evident from our GC/MS results that $S_{10\text{-}1}$ degraded cypermethrin by reductive dechlorination, oxidation or/and hydrolysis to transform to other 4 metabolites. The cypermethrin degradation pathway appeared to be different to the initial steps of SPs degradation by *Ochrobactrum lupine, Pseudomonas aeruginosa, Pseudomonas aeruginosa* and *Achromobacter* sp. [21, 22, 42, 43]. Moreover, no 3-phenoxybenzoic acid (3-PBA) was detected in the metabolites by GC-MS after 7 days of treatment, while 3-PBA was generally regarded as the major metabolite after hydrolysis of SPs in soil and water [21, 36, 42-44]. Owing to its antimicrobial activities [23, 45] and transient GC/MS detectable peak [21, 45], biodegradation of 3-PBA was rarely reported. Chen et al reported that fenvalerate was degraded by hydrolysis of the carboxylester linkage to yield 3-PBA, and then the intermediate was further utilized for bacterial growth by strain ZS-S-01, finally resulted in complete mineralization [42]. So, we speculated that carboxylesterases and oxidoreductases involved in degradation of cypermethrin by strain $S_{10\text{-}1}$, that needed to be testified by further experiments.

On the other hand, *R. sphaeroides* are metabolically flexible and under different situations they can grow chemoheterotrophically, chemoautotrophically, photoheterotrophically, and photoautotrophically [46]. Because of this multiplicity of growth modes there has been considerable interest in studying of degrading toxic compounds [35, 47]. The structural components of this metabolically diverse organism and their modes of integrated regulation are encoded by a genome of ~4.5 Mb in size [46]. Moreover, its large inventory of transport and chemotaxis genes also implies that *Rhodobacter* is adept at sensing and acquiring diverse compounds from its environment [48-50].

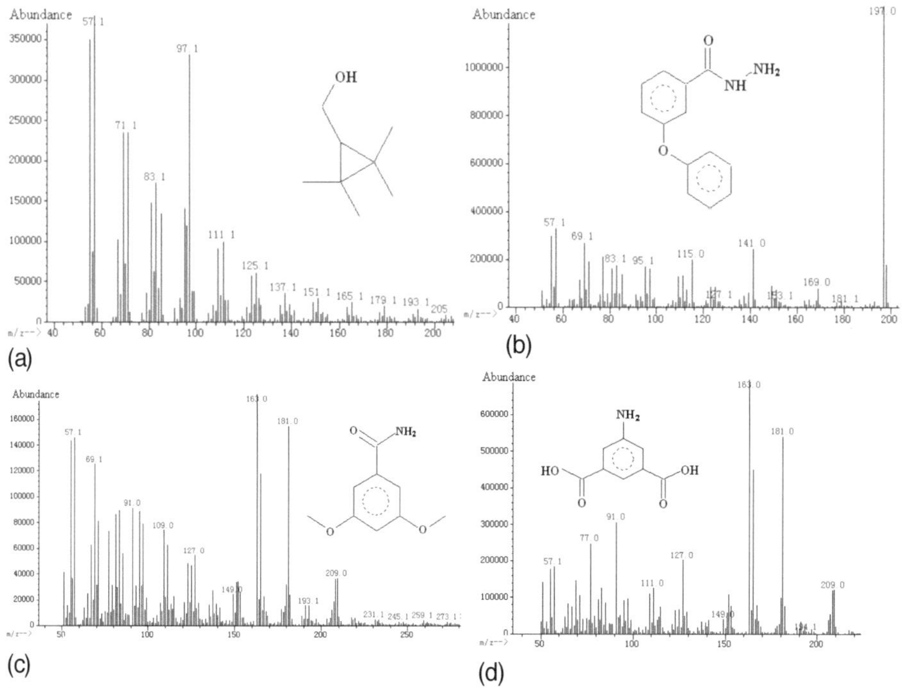

Figure 4. GC/MS spectra of four main metabolites produced during cypermethrin degradation by strain $S_{10\text{-}1}$. (a) cyclopropanemethanol; (b) 5-methoxy-2-nitrobenzoic acid; (c) 3,5-dimethoxybenzamide; (d) 5-aminoisophthalic acid.

4. Conclusion

R. sphaeroides strain $S_{10\text{-}1}$ was isolated from an activated sludge sample collected from the wastewater treatment pool of a pesticide plant. It can utilize cypermethrin as sole source of carbon, nitrogen and energy. The optimal temperature and pH for biodegradation of cypermethrin by strain $S_{10\text{-}1}$ were 35℃and pH 7.0, and the degradation rate reached 90.4% within 7 days under the optimal conditions. Four metabolic compounds were detected, hinting that there are complex redox reactions are involved in the cypermethrin degradation process.

In conclusion, our results indicated that strain $S_{10\text{-}1}$ could be a good choice for the bioremediation of cypermethrin contaminated water and soil. However, further studies such as its interactions with environment, toxicological aspects, degradation enzymes, biochemical and genetic aspects are still needed before the application in actual field-scale bioremediation.

Acknowledgements

This work was funded by the earmarked fund for Modern Agro-industry Technology Research System; the National Natural Science Foundation of China (No. 31071753), the key project of Hunan Provincial Education Department (11CY016), the key project of Shaoyang Municipal Science and Technology (J1107), and the 12th Five-Year key discipline of of Shaoyang University.

Author details

L. B. Yin[1], L. Z. Zhao[1], Y. Liu[2], D. Y. Zhang[2], S. B. Zhang[2] and K. Xiao[1]

1 Department of Biological and Chemical Engineering, Shaoyang University, Shaoyang, Hunan, People's Republic of China

2 Hunan Plant Protection Institute, Changsha, Hunan, People's Republic of China

References

[1] Vischetti, C, Monaci, E, Coppola, L, Marinozzi, M, & Casucci, C. Evaluation of biomassbed system in bio-cleaning water contaminated by fungicides applied in vineyard. International Journal of Environmental Analytical Chemistry. (2012). , 92(8), 949-962.

[2] Jaiswal, M, Chauhan, D, & Sankararamakrishnan, N. Copper chitosan nanocomposite: synthesis, characterization, and application in removal of organophosphorous pesticide from agricultural runoff. Environmental Science and Pollution Research. (2012). , 19(6), 2055-2062.

[3] Nkya, T. E, Akhouayri, I, Kisinza, W, & David, J. P. Impact of environment on mosquito response to pyrethroid insecticides: facts, evidences and prospects. Insect Biochemistry and Molecular Biology. http://dx.doi.org/10.1016/j.ibmb.(2012).

[4] Elsheikha, H. M, Mcorist, S, & Geary, T. G. Antiparasitic drugs: mechanisms of action and resistance. Essentials of Veterinary Parasitology; (2011).

[5] Ahn, K. C, Gee, S. J, Kim, H. J, Aronov, P. A, Vega, H, Krieger, R. I, & Hammock, B. D. Immunochemical analysis of 3-phenoxybenzoic acid, a biomarker of forestry

worker exposure to pyrethroid insecticides. Analytical and Bioanalytical Chemistry. (2011). , 401(4), 1285-1293.

[6] Nauen, R, Zimmer, C. T, Andrews, M, Slater, R, Bass, C, Ekbom, B, Gustafsson, G, Hansen, L. M, Kristensen, M, & Zebitz, C. P. W. Target-site resistance to pyrethroids in European populations of pollen beetle, Meligethes aeneus F.. Pesticide Biochemistry and Physiology. (2012). , 103(3), 173-180.

[7] Umina, P. A, Weeks, A. R, Roberts, J, & Jenkins, S. Peter Mangano, G., Lord, A., and Micic, S. The current status of pesticide resistance in Australian populations of the redlegged earth mite (Halotydeus destructor). Pest Management Science. (2012). , 68(6), 889-896.

[8] Ding, G, Shi, R, Gao, Y, Zhang, Y, Kamijima, M, Sakai, K, Wang, G, Feng, C, & Tian, Y. Pyrethroid pesticide exposure and risk of childhood acute lymphocytic leukemia in Shanghai. Environmental Science and Technology. (2012). DOI:es303362a.

[9] Soderlund, D. M. Molecular mechanisms of pyrethroid insecticide neurotoxicity: recent advances. Archives of Toxicology. (2012). , 86(2), 165-181.

[10] Alonso, M. B, Feo, M. L, Corcellas, C, Vidal, L. G, Bertozzi, C. P, Marigo, J, Secchi, E. R, Bassoi, M, Azevedo, A. F, & Dorneles, P. R. Pyrethroids: A new threat to marine mammals?. Environment International. (2012). , 47, 99-106.

[11] Tsuji, R, Yamada, T, & Kawamura, S. Mammal toxicology of synthetic pyrethroids. Topics in Current Chemistry. (2011). , 314, 83-111.

[12] Kulkarni, G, & Joshi, P. Cypermethrin and fenvalerate induced protein alterations in freshwater crab Barytelphusa cunicularis (westwood). Recent Research in Science and Technology. (2011). , 3(12), 7-10.

[13] Saha, S, & Kaviraj, A. Acute toxicity of synthetic pyrethroid cypermethrin to some freshwater organisms. Bulletin of Environmental Contamination and Toxicology. (2008). , 80(1), 49-52.

[14] Anderson, R. L. Toxicity of synthetic pyrethroids to freshwater invertebrates. Environmental Toxicology and Chemistry.(2009). , 8(5), 403-410.

[15] Datta, M, & Kaviraj, A. Acute toxicity of the synthetic pyrethroid pesticide fenvalerate to some air breathing fishes. Toxicological and Environmental Chemistry.(2011). , 93(10), 2034-2039.

[16] Anjum, R, Rahman, M, Masood, F, & Malik, A. Bioremediation of pesticides from soil and wastewater. Environmental Protection Strategies for Sustainable Development. (2012).

[17] Mugni, H, Demetrio, P, Bulus, G, Ronco, A, & Bonetto, C. Effect of aquatic vegetation on the persistence of cypermethrin toxicity in water. Bulletin of Environmental Contamination and Toxicology.(2011). , 86(1), 23-27.

[18] Riser-roberts, E. Remediation of petroleum contaminated soils: biological, physical, and chemical processes: (CRC); (1998).

[19] Marttinen, S, Kettunen, R, Sormunen, K, Soimasuo, R, & Rintala, J. Screening of physical-chemical methods for removal of organic material, nitrogen and toxicity from low strength landfill leachates. Chemosphere. (2002). , 46(6), 851-858.

[20] Naveen, D, Majumder, C, Mondal, P, & Shubha, D. Biological treatment of cyanide containing wastewater. Research Journal of Chemical Sciences. (2011). , 1(7), 15-21.

[21] Chen, S, Hu, M, Liu, J, Zhong, G, Yang, L, Rizwan-ul-haq, M, & Han, H. Biodegradation of beta-cypermethrin and 3-phenoxybenzoic acid by a novel Ochrobactrum lupini DG-S-01. Journal of Hazardous Materials. (2011). , 187(1), 433-440.

[22] Zhang, C, Wang, S, & Yan, Y. Isomerization and biodegradation of beta-cypermethrin by Pseudomonas aeruginosa CH7 with biosurfactant production. Bioresource technology. (2011). , 102(14), 7139-7146.

[23] Chen, S, Geng, P, Xiao, Y, & Hu, M. Bioremediation of β-cypermethrin and 3-phenoxybenzaldehyde contaminated soils using Streptomyces aureus HP-S-01. Applied Microbiology and Biotechnology. (2012). , 94(2), 505-515.

[24] Kontur, W. S, Ziegelhoffer, E. C, Spero, M. A, Imam, S, Noguera, D. R, & Donohue, T. J. Pathways involved in reductant distribution during photobiological H_2 production by Rhodobacter sphaeroides. Applied and Environmental Microbiology.(2011). , 77(20), 7425-7429.

[25] Sonhom, R, Thepsithar, C, & Jongsareejit, B. High level production of 5-aminolevulinic acid by Propionibacterium acidipropionici grown in a low-cost medium. Biotechnology letters. (2012). , 34(9), 1667-1672.

[26] Kang, Z, Wang, Y, Wang, Q, & Qi, Q. Metabolic engineering to improve 5-aminolevulinic acid production. Bioengineered. (2011). , 2(6), 342-345.

[27] Yin, L, Li, X, Liu, Y, Zhang, D, Zhang, S, & Luo, X. Biodegradation of cypermethrin by Rhodopseudomonas palustris GJ-22 isolated from activated sludge. Fresenius Environmental Bulletin. (2012). , 21(2), 397-405.

[28] Mirnejad, R, Babavalian, H, Moghaddam, M. M, Khodi, S, & Shakeri, F. Rapid DNA extraction of bacterial genome using laundry detergents and assessment of the efficiency of DNA in downstream process using polymerase chain reaction. African Journal of Biotechnology. (2012). , 11(1), 173-178.

[29] Haws, D. C, Hodge, T. L, & Yoshida, R. Optimality of the neighbor joining algorithm and faces of the balanced minimum evolution polytope. Bulletin of Mathematical Biology. (2011). , 73(11), 2627-2648.

[30] Liu, S, Yao, K, Jia, D, Zhao, N, Lai, W, & Yuan, H. A pretreatment method for HPLC analysis of cypermethrin in microbial degradation systems. Journal of Chromatographic Science. (2012). , 50(6), 469-476.

[31] Vos, P, Garrity, G, Jones, D, Krieg, N. R, Ludwig, W, Rainey, F. A, Schleifer, K. H, & Whitman, W. B. Bergey's Manual of Systematic Bacteriology: The Firmicutes, Volume 3, (Springer); (2009). , 3

[32] Ramana, C. V. Photoassimilation of aromatic compounds by Rhodobacter sphaeroides OU5. Ph.D Thesis. University of Hyderabad Idia; (2012).

[33] Yetis, M, Gündüz, U, Eroglu, I, Yücel, M, & Türker, L. Photoproduction of hydrogen from sugar refinery wastewater by Rhodobacter sphaeroides OU001. International Journal of Hydrogen Energy.(2000). , 25(11), 1035-1041.

[34] Tao, Y, He, Y, Wu, Y, Liu, F, Li, X, Zong, W, & Zhou, Z. (2008). Characteristics of a new photosynthetic bacterial strain for hydrogen production and its application in wastewater treatment. International Journal of Hydrogen Energy.2008;, 33(3), 963-973.

[35] Chalam, A, Sasikala, C, & Ramana, C. V. and Raghuveer Rao, P. Effect of pesticides on hydrogen metabolism of Rhodobacter sphaeroides and Rhodopseudomonas palustris. FEMS Microbiology Ecology.(1996). , 19(1), 1-4.

[36] Lin, Q, Chen, S, Hu, M, Rizwan-ul-haq, M, Yang, L, & Li, H. (2011). Biodegradation of cypermethrin by a newly isolated actinomycetes HU-S-01 from wastewater sludge. International Journal of Environment and Science and Technology. 2011;, 8(1), 45-56.

[37] Howe, G. E, Marking, L. L, Bills, T. D, Rach, J. J, & Mayer, F. L. Effects of water temperature and pH on toxicity of terbufos, trichlorfon, 4-nitrophenol and 2,4-dinitrophenol to the amphipod Gammarus pseudolimnaeus and rainbow trout (Oncorhynchus mykiss). Environmental Toxicology and Chemistry.(2009). , 13(1), 51-66.

[38] Rowland, S, Jones, D, Scarlett, A, West, C, Hin, L. P, Boberek, M, Tonkin, A, Smith, B, & Whitby, C. Synthesis and toxicity of some metabolites of the microbial degradation of synthetic naphthenic acids. Science of the Total Environment. (2011). , 409(15), 2936-2941.

[39] Wang, B, Ma, Y, Zhou, W, Zheng, J, Zhu, J, He, J, & Li, S. Biodegradation of synthetic pyrethroids by Ochrobactrum tritici strain pyd-1. World Journal of Microbiology and Biotechnology.(2011). , 27(10), 2315-2324.

[40] Zhang, S, Yin, L, Liu, Y, Zhang, D, Luo, X, Cheng, J, Cheng, F, & Dai, J. Cometabolic biotransformation of fenpropathrin by Clostridium species strain ZP3. Biodegradation.(2011). , 22(5), 869-875.

[41] Tallur, P. N, Megadi, V. B, & Ninnekar, H. Z. Biodegradation of cypermethrin by Micrococcus sp. strain CPN 1. Biodegradation.(2008). , 19(1), 77-82.

[42] Chen, S, Yang, L, Hu, M, & Liu, J. Biodegradation of fenvalerate and 3-phenoxybenzoic acid by a novel Stenotrophomonas sp. strain ZS-S-01 and its use in bioremediation of contaminated soils. Applied Microbiology and Biotechnology.(2011). , 90(2), 755-767.

[43] Chen, S, Zhang, Y, Hu, M, Geng, P, Li, Y, & An, G. Bioremediation of β-cypermethrin and 3-phenoxybenzoic acid in soils. In Water Resource and Environmental Protection (ISWREP), International Symposium on, IEEE), (2011). , 3, 1717-1721.

[44] Mccoy, M. R, Yang, Z, Fu, X, Ahn, K. C, Gee, S. J, Bom, D. C, Zhong, P, Chang, D, & Hammock, B. D. Monitoring of total type II pyrethroid pesticides in Citrus oils and water by converting to a common product phenoxybenzoic acid. Journal of Agricultural and Food Chemistry.(2012). , 3.

[45] Chen, S, Hu, Q, Hu, M, Luo, J, Weng, Q, & Lai, K. Isolation and characterization of a fungus able to degrade pyrethroids and phenoxybenzaldehyde. Bioresource Technology.(2011). , 3.

[46] Mackenzie, C, Choudhary, M, Larimer, F. W, Predki, P. F, Stilwagen, S, Armitage, J. P, Barber, R. D, Donohue, T. J, Hosler, J. P, & Newman, J. E. The home stretch, a first analysis of the nearly completed genome of Rhodobacter sphaeroides 2.4. 1. Photosynthesis Research.(2001). , 70(1), 19-41.

[47] Barber, R. D, & Donohue, T. J. Function of a glutathione-dependent formaldehyde dehydrogenase in Rhodobacter sphaeroides formaldehyde oxidation and assimilation. Biochemistry.(1998). , 37(2), 530-537.

[48] Armitage, J. P, & Schmitt, R. Bacterial chemotaxis: Rhodobacter sphaeroides and Sinorhizobium meliloti variations on a theme? Microbiology Reading.(1997). , 143, 3671-3682.

[49] Porter, S. L, Wadhams, G. H, & Armitage, J. P. Signal processing in complex chemotaxis pathways. Nature Reviews Microbiology.(2011). , 9(3), 153-165.

[50] Kojadinovic, M, Sirinelli, A, Wadhams, G. H, & Armitage, J. P. New motion analysis system for characterization of the chemosensory response kinetics of Rhodobacter sphaeroides under different growth conditions. Applied and Environmental Microbiology.(2011). , 77(12), 4082-4088.

Chapter 2

Chlorophyll Biodegradation

Nina Djapic

Additional information is available at the end of the chapter

http://dx.doi.org/10.5772/56155

1. Introduction

One of the most important biochemical processes is photosynthesis and life depends on the conversion of the solar energy into a chemical one. It occurs in photoautotrophic organisms. The chlorophyll *a* is supported by chlorophyll *b*, carotenes and xanthophylls in the light capture. The chlorophyll *a* and accessory pigments are found in cell organelles called chloroplasts. The conversion of light energy into the chemical one can be described thermodynamically. The entropy of photosynthesis yields the energy of 27.32 kJ mol^{-1} and is a source of energy for the plants. The atmospheric oxygen is a product of the oxygenic photosynthesis, the product of the first phase process known as light – dependant reactions. In those reactions, the chlorophylls, in the so called reaction centre, function as an absorbent of a photon initiating the electron transport chain, with the loss of an electron. During the electron flow the energy is produced and the reduced form of the nicotinamide adenine dinucleotide phosphate (NADPH) and the adenosine triphosphate (ATP) are formed. The chlorophyll molecule regains the electron from water. The other phase of photosynthesis is called the light – independent reactions or the Calvin cycle. In those reactions the enzyme ribulose-1, 5-bisphosphate carboxylase/oxygenase (RuBisCO) captures the CO_2 from the atmosphere and during the Calvin cycle glucose is formed. The light-absorbing green pigments, the chlorophylls, are biodegraded under various conditions: with the leaf senescence in autumn, under stress conditions, such as, drought, flooding, chilling, mechanical wounding, etc., under the exposure to the ethylene and after the treatment of plants with herbicides. In autumn deciduous trees shed tones of leaves. The green leaves change the colour due to the chlorophyll biodegradation. The peculiarity of the chlorophyll biodegradation in autumnal leaves will be described.

2. Chlorophyll biodegradation

The chlorophyll *a* and chlorophyll *b* biodegradation pathways consist of a great number of steps and up to now only the steps where the chlorophylls' biodegradation products have a chromophore that can absorb the ultra-violet (UV)/visible light are known. Each step can be considered separately. In the cell, every step is coordinated, highly regulated and most steps are enzymatically catalyzed. The first group of reactions includes the loss of magnesium, the loss of phytol and the modifications on the periphery of the chlorophyll nucleus in which the aromatic tetrapyrrole macrocycle remains intact [1]. The starting investigations on chlorophylls and chlorophylls' biodegradation products were done by Richard Willstaetter and Arthur Stoll who have discovered the nature of the chlorophylls [2]. Ivan Parfen'ievitch Borodin was the first who obtained crystalline chlorophyll [2]. Finally, Mikhail Semenovitch Tswett, after the formulation of the chromatography method, proved that the chlorophyll was a mixture of two green pigments "chlorophyllins alfa and beta" and that Ivan Parfen'ievitch Borodin's pigment is different from natural chlorophylls [3, 4, 5, 6]. Richard Willstaetter and Arthur Stoll showed that green leaves contain an enzyme chlorophyllase which hydrolyses the phytyl ester of the chlorophylls' propionate group to yield chlorophyllide *a* [2, 7, 8]. In the presence of alcohol, the crystalline substance formed by transesterification, is methyl chlorophyllide [9]. Richard Willstaetter and Arthur Stoll determined the basic structure of chlorophyll *a* and *b* and showed that Ivan Parfen'ievitch Borodin's crystalline pigments were chlorophyllide *a* and *b* formed from natural chlorophylls by the hydrolysis by the enzyme chlorophyllase [10].

The interconversion of chlorophyll *b* to chlorophyll *a* and vice versa occurs in oxygenic photosynthetic organisms [11]. The suggested chlorophyll cycle is depicted in Figure 1. The cycle starts with the reduction of chlorophyll *b* by chlorophyll *b* reductase, an NADPH dependant enzyme, to form 7^1–hydroxyl chlorophyll *a*. The 7^1–hydroxyl chlorophyll *a* is a stable intermediate and it was isolated from higher plants. The next reduction step is catalyzed by ferredoxin dependant 7^1–hydroxyl chlorophyll *a* reductase to form chlorophyll *a*. It is also suggested that chlorophyll *a* interconverts to chlorophyll *b* by an oxygen dependant enzyme chlorophyll *a* oxygenase. The enzyme oxidizes chlorophyll *a* to 7^1–hydroxyl chlorophyll *a* which is further oxidized to chlorophyll *b* by 7^1–hydroxyl chlorophyll *a* dehydrogenase.

The bioconversion of chlorophyll *b* to chlorophyll *a* undergoes through a 7^1–hydroxyl intermediate. The chlorophyll interconversion cycle can be expanded to the following catabolic steps toward the chlorophyllide *a* (Figure 1). The chlorophyll *a* is enzymatically hydrolysed to chlorophyllide *a*. The chlorophyll *b* is enzymatically hydrolysed to chlorophyllide *b* and undergoes through the formation of 7^1–hydroxyl chlorophyllide *a* which is enzymatically reduced to the chlorophyllide *a* [12].

Richard Willstaetter observed that the ash from the crude chlorophylls is rich in magnesia [2]. The magnesium is cleaved from the chlorophylls by a treatment with mild acids. The magnesium coordination bond is stable against alkali [9]. On the other hand, treatment with hydroxide saponifies the chlorophylls' phytol side chain yielding chlorophyllides.

http://dx.doi.org/10.5772/56155

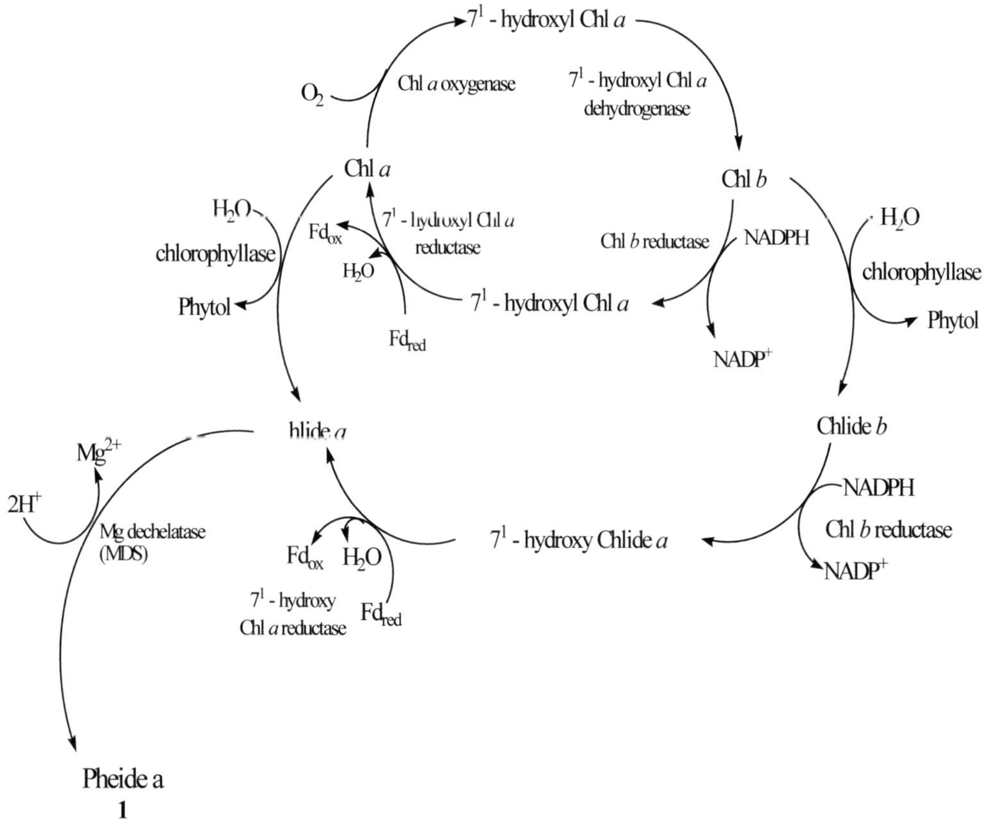

Figure 1. The chlorophyll cycle

The chlorophyll biodegradation continues with dechelating the chlorophyllide *a* molecule. The enzyme involved in demetallation is Mg–dechelatase. A catalytic cofactor called Mg–dechelating substance (MDS) is associated with the activity of Mg–dechelatase [13]. When the magnesium is expelled from the core of chlorophyllide *a*, the remaining structure is called pheophorbide *a* (1) (Figure 1).

The next group of reactions involves the cleavage of the macrocyclic aromatic ring system yielding open chain tetrapyrrolic compound, the *seco*–phytoporhyrins, in which side chains can be further modified. In further chlorophyll biodegradation, pheophorbide *a* (1) methene bridge at the C4–C5 position is oxidized by the pheophorbide *a* oxygenase (P*a*O) to yield the 4,5-dioxo-*seco*-pheophorbide *a* chlorophyll biodegradation product (2). The P*a*O is an iron–dependant monooxygenase. The electrons required for the redox cycle are supplied from the reduced ferredoxin (Figure 2).

The proposed mechanism is based on the single oxygen attachment to the double bond at the C4=C5 position forming regioselective intermediate by the cleavage of the oxirane ring.

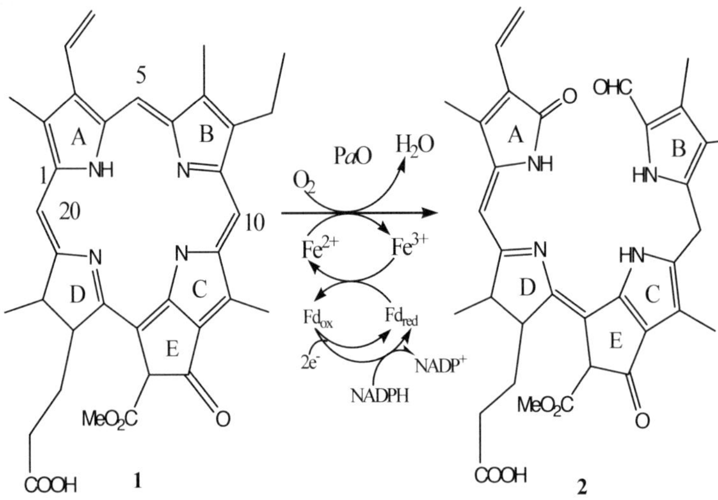

Figure 2. The proposed mechanism of the pheophorbide *a* (1) oxidation, by the PaO to the 4,5-dioxo- *seco*-pheophorbide *a* (2).

Subsequently, the oxiran ring is opened hydrolytically. The retro–aldol reaction proceeds and the rearrangement of protons form the 4,5-dioxo-*seco*-pheophorbide *a* (2) (Figure 2).

The amount of accumulated chlorophyll biodegradation products within the same plant species varies. The quantity of the chlorophyll biodegradation products in senescent leaves most probably depends on the time of the collection, seasonal climate, developmental stage of the plant, enzymes, etc. One of the reasons for the chlorophyll biodegradation in biological systems is due to the tendency of the organism to decrease the level of photodynamically active chlorophyll before the programmed cell death [14].

3. Chlorophyll biodegradation in Hamamelidaceae and Vitaceae autumnal leaves

The autonomous induction of leaf senescence occurs in autumn. The major consequence of leaf senescence is chlorophyll biodegradation. The crude methanol extracts of Hamamelidaceae and Vitaceae autumnal leaves' plant species were analyzed on the reversed phase (RP): RP–C_8 and RP–C_4 analytical columns by liquid chromatography–mass spectrometry (LC–MS) and their structure was determined by mass and nuclear magnetic resonance (NMR) spectra [15, 16, 17]. The results obtained gave insight in the chlorophyll biodegradation in plant species of Hamamelidaceae and Vitaceae autumnal leaves (Figure 3). The chlorophyll biodegradation pathway continues from the 4, 5-dioxo-*seco*-pheophorbide *a* (2) to the formation of the hydroxylated ethyl side chain chlorophyll biodegradation product (3). The enzyme catalyzing the hydroxylation of the ethyl side chain is still unknown.

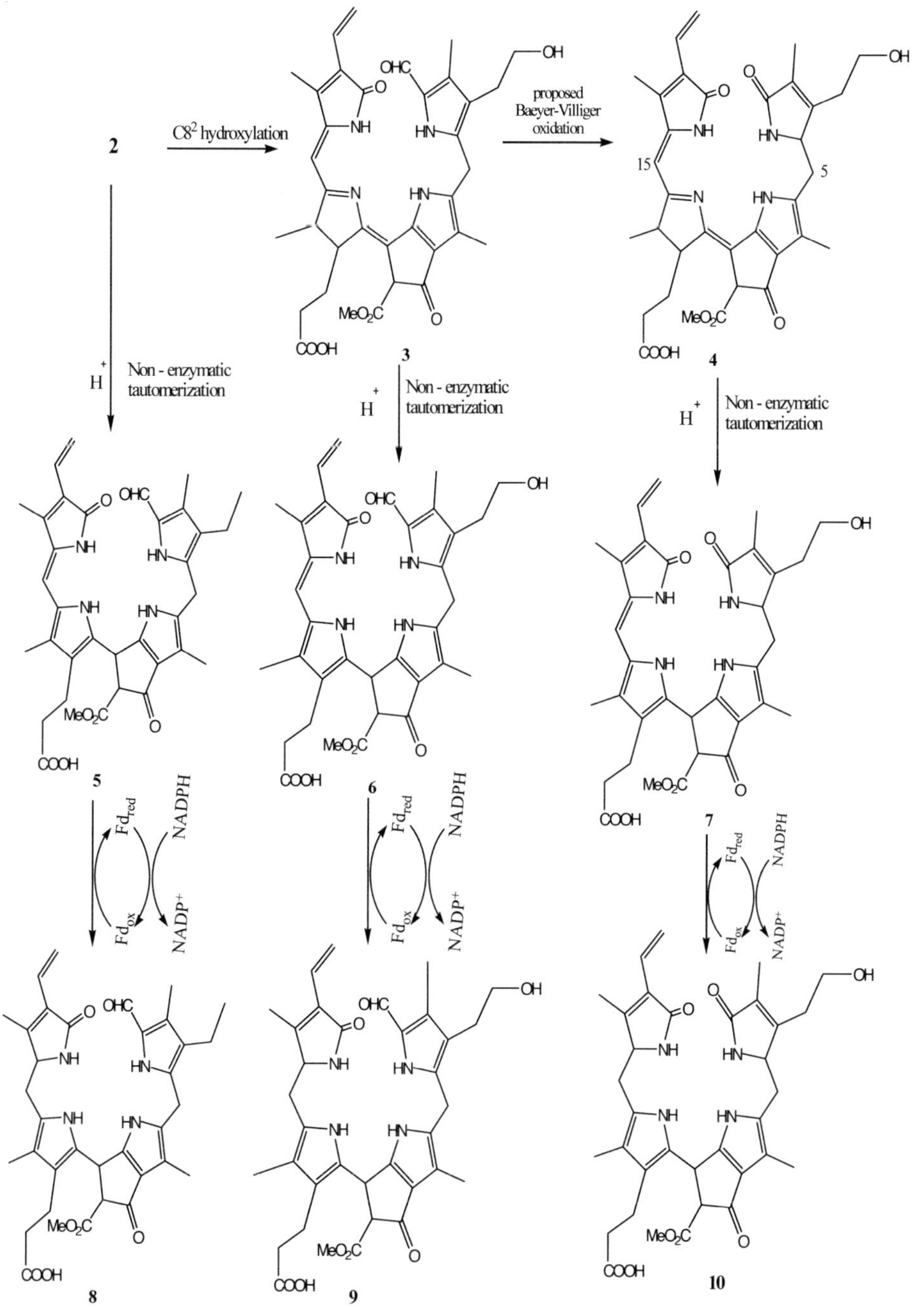

Figure 3. The chlorophyll biodegradation pathway in Hamamelidaceae and Vitaceae autumnal leaves. The ethyl side chain of the 4, 5-dioxo-*seco*-pheophorbide *a* (2) is enzymatically hydroxylated to the compound 3 and after the proposed Baeyer-Villiger oxidation the urobilinogenic chlorophyll biodegradation product (4) is formed. All chlorophyll biodegradaton products under acid conditions form the thermodynamically more stable compounds (5, 6, 7). The reduction of the chlorophyll biodegradation compounds 5, 6 and 7 proceeds via the reduction of the "western" methene bridge forming the chlorophyll biodegradation products 8, 9 and 10.

The urobilinogenic chlorophyll biodegradaton products isolated from autumnal leaves of *Parrotia persica* and *Hamamelis virginiana*, Hamamelidaceae significantly differ from other isolated chlorophyll biodegradation products in one functional group [15, 16]. The aldehyde group at C5 position is absent and chlorophyll biodegradation product's structure refers to the urobilinogen. The aldehyde group at the C5 (3) position might be oxidized by Baeyer–Villiger monooxygenase. The carbon numeration differs from all previously isolated chlorophyll biodegradation products and is as in urobilinogen (Figure 3) [15, 16].

The proposed oxidation step is by the Baeyer–Villiger monooxygenase, from the C8^2-hydroxylated chlorophyll biodegradation product (3) to the urobilinogenic chlorophyll biodegradation product (4). The oxidation results in the formation of an additional chiral centre at the position C4. The absolute configuration of all chiral centres is unknown. It is proposed that this enzyme can convert the chlorophyll biodegradation product's aldehyde group into the corresponding dehydro pyrrolidin–2–one of the urobilinogenic chlorophyll biodegradation product mediated by the hydroperoxide intermediate. The nucleophilic addition of the hydroperoxide to the C5 oxo group affords peroxide intermediate. After the rearrangement, the insertion of the oxygen is made. After the ester hydrolysis, the alcohol is formed [17, 18, 19]. After the acidic catalyzed tautomerization, the urobilinogenic chlorophyll biodegradation product is formed. Another conceivable mechanism that can explain the formation of the urobilinogenic chlorophyll biodegradation product is by the oxidative decarboxylation of an acid intermediate generated by the oxidation of the aldehyde group.

The chlorophyll biodegradation products (2, 3, 4) form the thermodynamically more stable compounds (5, 6, 7) after the acidic catalyzed tautomerization, [20]. The reduction of the compounds 5, 6 and 7 proceed via the reduction of the "western" methene bridge. It is catalyzed by a reductase. The reductase is a ferredoxin dependant enzyme with no requirements of cofactors such as flavin or metals. After the reduction of the chlorophyll biodegradation products 5, 6, 7 the chlorophyll biodegradation products 8, 9 and 10 are formed.

Further chlorophyll biodegradation products were, up to now, not detected. The further chlorophyll biodegradation products lose the chromophore that can absorb the UV light, they become colourless, and were not identified up to now.

4. Chlorophyll biodegradation in Solanaceae autumnal leaves

In autumnal leaveas of the Solanaceae family: tobacco (*Nicotiana tabacum*), potatoe (*Solanum tuberosum*), *Atropa belladonna*, etc. chlorophyll biodegradation products are present. In Solanaceae plant species the 4, 5-dioxo-*seco*-pheophorbide *a* (2) precedes, in further chlorophyll biodegradation, the hydroxylation at the C8^2 position and then the glycosylation at the C8^2 position (11) [21]. The glycosyl bond is most probably introduced by glycosyltransferase. The Solanaceae chlorophyll biodegradation product (11) form a thermodynamically stable aromatic ring D under the acidic conditions (12) (Figure 4). The reduction of the thermodynamically more stable compound (12) proceeds via the reduction of the "western" methene

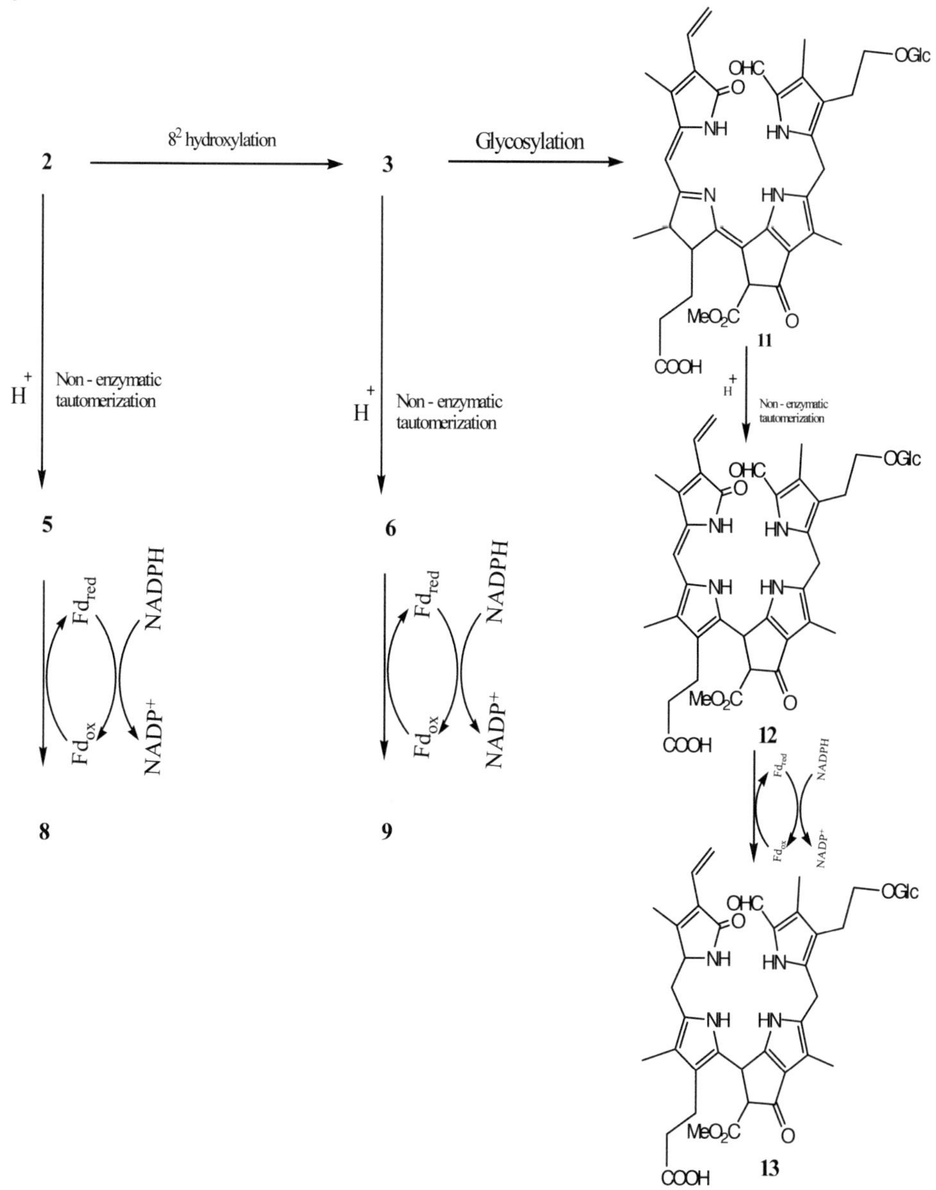

Figure 4. The chlorophyll biodegradation pathway in Solanaceae autumnal leaves, known up to now.

bridge. The reductase catalyzed reaction proceeds and the glycosylated chlorophyll biodegradation product 13 is formed.

The detection of other chlorophyll biodegradation products present in Solanaceae autumnal leaves will extend the chlorophyll biodegradation pathway, known up to now. The isolation of the sufficient quantity of the Solanaceae glycosylated chlorophyll biodegradation products

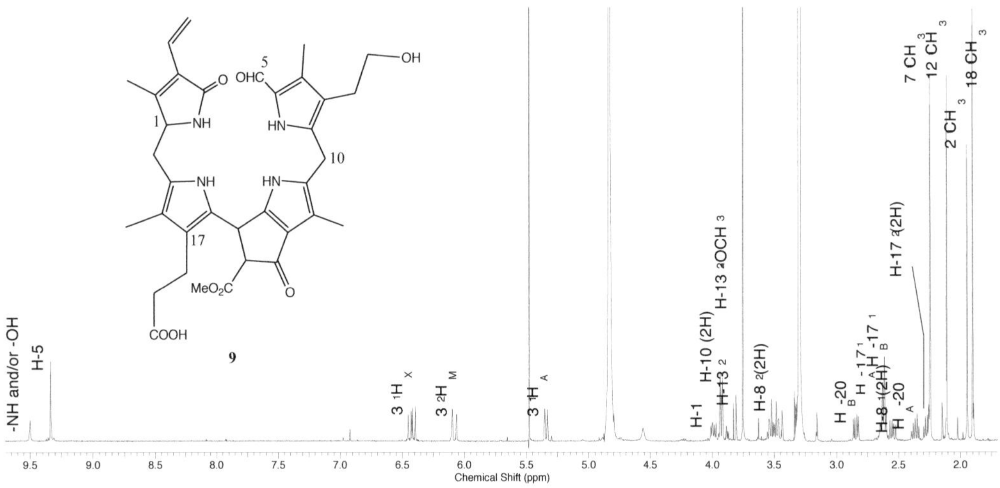

Figure 5. The proton spectrum of the compound 9.

will provide the information about the conformation of the sugar part and the information regarding the linkage between the aglycone and the sugar, such as, in the case of the anthocynins [22, 23, 24, 25].

5. Chlorophyll biodegradation in *Ginkgo biloba* autumnal leaves

Crude methanol extract of autumnal leaves' of *Ginkgo biloba*, Ginkgoaceae male plant were analyzed by LC-ESI-MS and LC/ESI-MS chromatograms and spectra obtained revealed the presence of the chlorophyll biodegradation products. Of hundreds of samples analyzed, only in one chromatogram obtained the traces the chlorophyll biodegradation product 9 with the molecular ion at *m/z* 645 for the molecular formula $C_{35}H_{40}N_4O_8$ was observed. Its proton NMR spectrum is depicted in Figure 5. Integration of the proton NMR spectrum revealed the presence of 33 protons. The signal of one proton on the heteroatom was still not exchanged with deuterium and was present in the low field region. The aldehyde H-5 proton signal was present in the low field region. Three low field *doublet of doublets* sets in the spectrum indicated the presence of an AMX spin-splitting pattern. The proton signal H-15 was overlapped with the HDO signal. Propionyl side chain H-17^1 protons came in a narrow δ range along with the methylene bridge H-20 protons. The H-17^2 protons, next to the electron withdrawing carboxyl group, had chemical shifts in the upfield region compared to H-17^1 protons. The most prominent peak in the proton spectrum was the signal assigned to the methoxycarbonyl group at the position 8^2. The methyl groups and other protons were assigned from the "1D difference" NOE spectra. In the COSY spectrum, the well resolved downfield 3^2 H_M vinyl proton was assigned to the geminal 3^2 H_A proton tracing further to the vicinal 3^1 H_X proton. This vinyl isolated spin system terminated at the 3^1 H_X proton.

All chromatograms obtained revealed the presence of three peaks, the UV detection at λ=312 nm. One peak had no Gaussian peak resolution (*Gb*-1), other two peaks coeluted with no separation to the baseline (*Gb*-2 and *Gb*-3) at the 25°C with gradient elution water (0.1% TFA): methanol on RP C_4 – analytical column (Figure 6). All peaks had the *m*/*z* 635. The ESIMS of the *Gb*-3 eluting at the 37.8 min. is depicted in Figure 7.

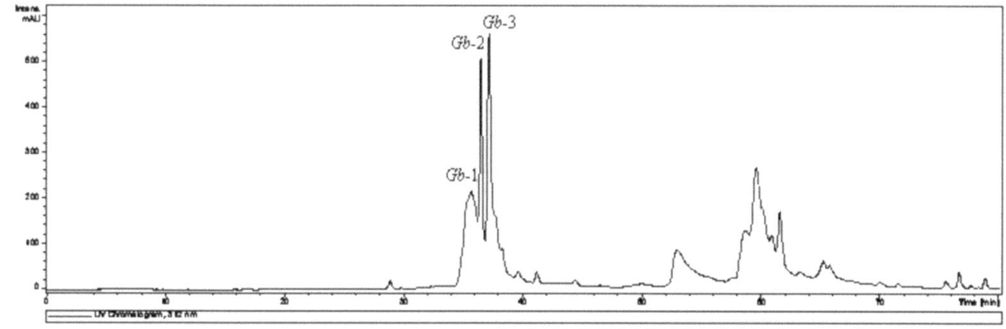

Figure 6. The chromatogram of *Gingko biloba* male methanol autumnal leaves' crude extract. The LC conditions: Column: Nucleosil 100-5 C_4 4x250 mm. The mobile phase: 90% v/v water (0.1% TFA):methanol to 0% v/v water (0.1%TFA):methanol in 80 minutes. Flow rate: 0.2 ml/min. UV detection at λ=312.

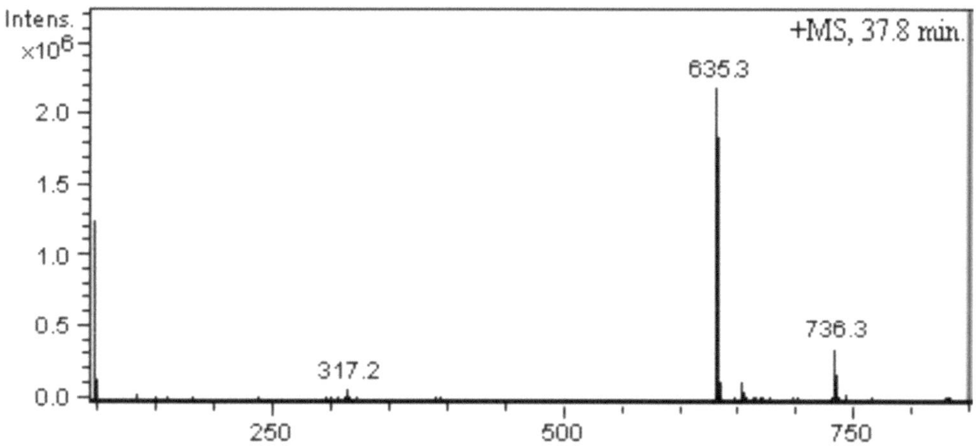

Figure 7. The ESIMS of the *Gb*-3 eluting at the 37.8 min.

The semi-preparative separation on one HPLC instrument on the semi-preparative RP-C_4 column with water (0.1% TFA): methanol solvent mixture gradient elution revealed the chromatogram (extracted at λ=312 nm) with complete peak distortion of *Ginkgo biloba* chlorophyll biodegradation products (Figure 8). The peak distortion was probably due to compounds' secondary equilibrium. It appears that the conversion rate of *Ginkgo biloba* compounds increased during the separation under the applied separation conditions. The

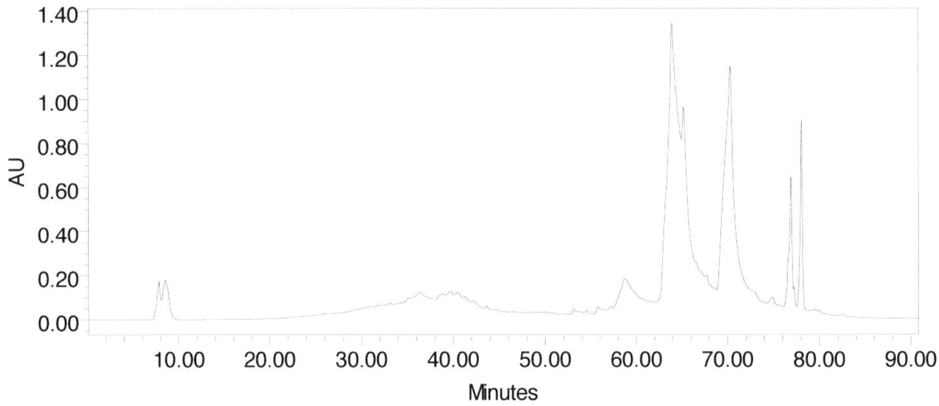

Figure 8. The chromatogram of *Ginkgo biloba* autumnal leaves, UV detection at λ=312 nm when the separation was done on one older model of the HPLC instrument

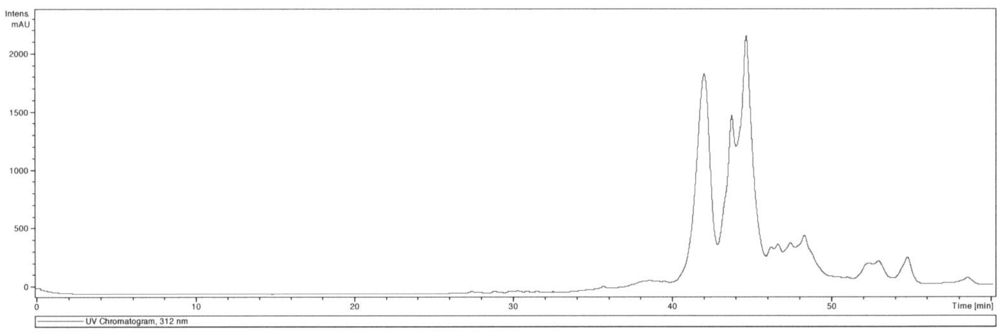

Figure 9. The chromatogram of *Ginkgo biloba* leaves' methanol crude extract. The LC conditions: Column: Nucleosil 120-7 C_4 10x250 mm. The mobile phase: 90% v/v water (acetate buffer): 10% v/v methanol to 0% v/v water (acetate buffer): 100% v/v methanol at 22°C in 90 minutes. Flow rate: 1.2 ml/min. UV detection at 312 nm.

peaks became smaller spreading over a long elution time (Figure 8). The UV spectrum of the broad peak from 40 to 42 min. revealed the presence of a chromophore with the absorption band of 317 nm. The *Ginkgo biloba* chlorophyll biodegradation products' peaks were broad and highly distorted.

The separation investigations continued on semi-preparative RP C_4 using the latest model of one LC instrument. The solvent mixture used was water (acetate buffer (pH 3.75)): methanol. The temperature was changed (Figure 9 and 10). At 30°C the separation between the peaks was observed (Figure 10). The isolation of *Ginkgo biloba* chlorophyll biodegradation products was done as depicted in Figure 10. Three collected fractions were evaporated under reduced pressure (t<40°C) and MS and NMR spectra were recorded.

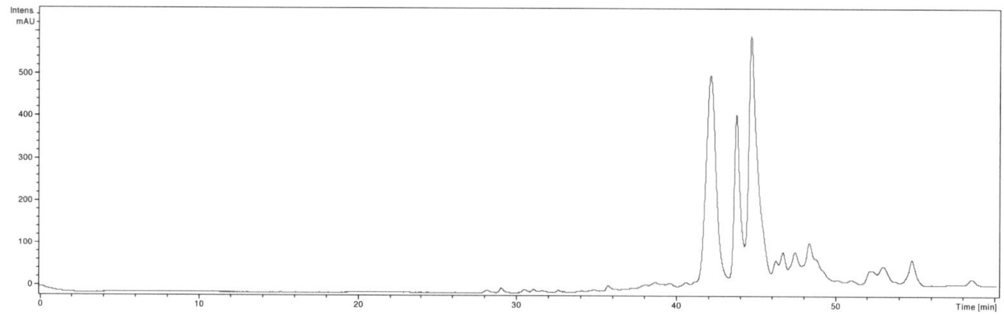

Figure 10. The chromatogram of *Ginkgo biloba* leaves' methanol crude extract. The LC conditions: Column: Nucleosil 120-7 C_4 10x250 mm. The mobile phase: 90% v/v water (acetate buffer): 10% v/v methanol to 0% v/v water (acetate buffer): 100% v/v methanol at 30°C in 90 minutes. Flow rate: 1.2 ml/min. UV detection at 312 nm.

The compound eluting at 42 min. is assigned as the *Gb*-1, at the 44 min. as the *Gb*-2 and at the 45 min. as the *Gb*-3. The determination of *Ginkgo biloba* chlorophyll biodegradation products was done by mass spectrometry and NMR spectroscopy. The High Resolution Electron Spray Ionisation Mass Spectrometry (HRESIMS) of the *Gb*-1 showed a molecular ion at *m/z* 679.2350 for the molecular formula $C_{33}H_{37}N_4O_9Na_2$ $[M+2Na]^{2+}$, calculated *m /z* 679.2350, Δ 0.00 ppm. The other ion had *m/z* 657.2530 for the molecular formula $C_{33}H_{37}N_4O_9Na$ $[M+Na]^{+}$, calculated *m /z* 657.2531, Δ + 0.15 ppm. The formic acid was added to the sample and the HRESIMS was recorded. The mass spectrum obtained revealed the presence of *m/z* 679.2351 and *m/z* 657.2530. The *Gb*-2 had a molecular ion at *m/z* 679.2350 for the molecular formula $C_{33}H_{37}N_4O_9Na_2$ $[M+2Na]^{2+}$, calculated *m /z* 679.2750, Δ 0.00 ppm. The other molecular ion had *m/z* 701.2172 for the molecular formula $C_{33}H_{36}N_4O_9Na_3$ $[M+3Na]^{3+}$, calculated *m/z* 701.2170, Δ– 0.15 ppm. The addition of formic acid to the sample revealed the presence of the peaks *m/z* 657.2531, *m/z* 679.2347 and *m/z* 701.2166. For the *Gb*-3 compound, the Electron Spray Ionisation Mass Spectrometry (ESIMS) revealed the presence of three peaks, *m /z* 679.24, 701.22 and 723.20 when the sample was dissolved in methanol. The sample dissolved in methanol with formic acid revealed the presence of three peaks *m /z* 657.25, 679.24 and 701.22.

The proton NMR spectra were recorded. In the *Gb*-1 proton spectrum signals for the aldehyde proton were present, two well resolved *singlets* and one *pseudo triplet* (Figure 11). The *Gb*-2 and *Gb*-3 had the same signals in the low field (Figure 12 and 13).

The formyl group can have two conformations NO – *Z* (14) and NO – *E* (15) (Figure 14) [26, 27]. In all previously isolated *seco* – phytoporphyrins such conformational change has not been observed.

The vinyl group proton signals were of unresolved line shape (Figure 11, 12 and 13). The *doublet of doublets* were fairly broad at the room temperature. The vicinal proton of the vinyl group showed two superpositioned *doublets of doublets*. Similar to the proton spectrum of the chlorophyll biodegradation product 9, in the low field, three *doublet of doublets* sets in the proton spectrum indicated the presence of an AMX spin-splitting pattern (Figure 5, 11, 12 and 13). In the low field two sharp *singlets* were recorded (Figure 11). They were or impurities or the

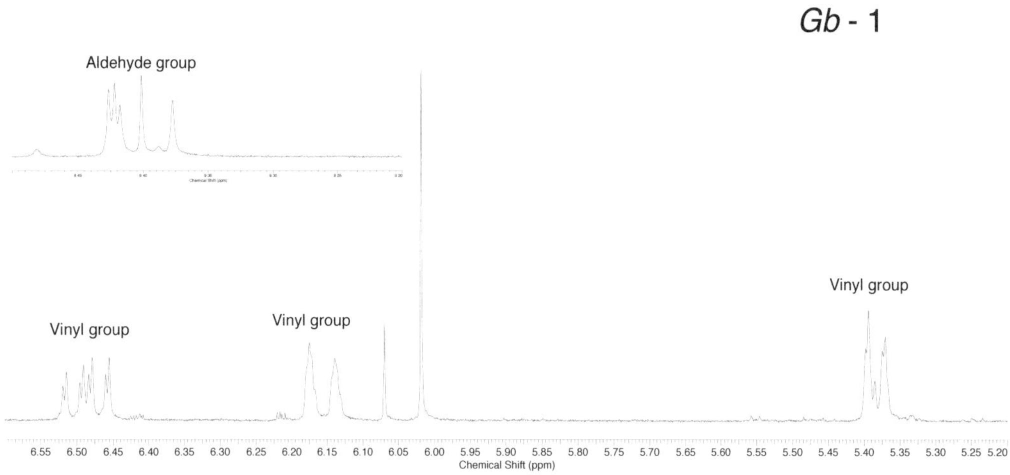

Figure 11. The enlarged proton spectrum of *Gb*-1 in the low field region

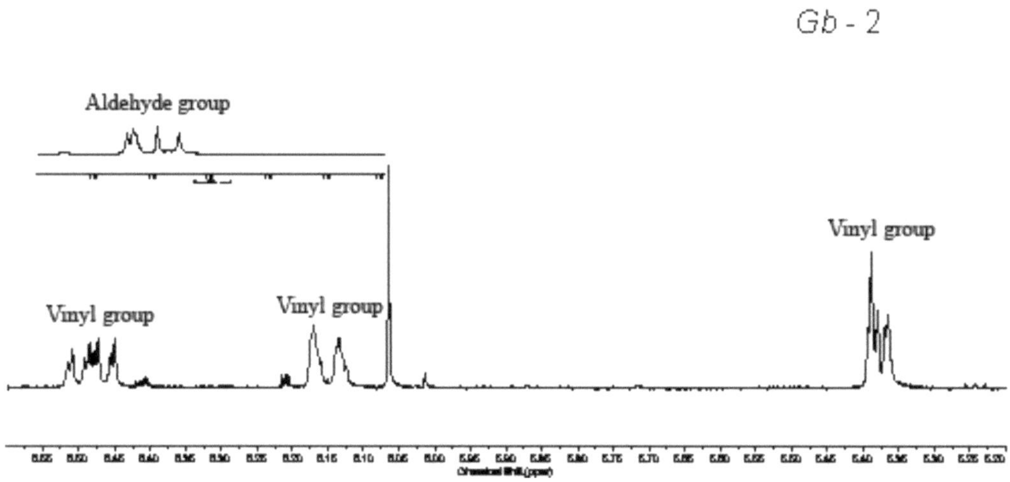

Figure 12. The enlarged proton spectrum of *Gb*-2 in the low field region

proton at the "western" methylene bridge. In the *Gb*-2 chlorophyll biodegradation product's proton spectrum in the vinyl region the proposed "western" methylene bridge proton gave one *singlet* (Figure 12). In the *Gb*-3 compound's low field proton spectrum only vinyl signals were recorded (Figure 13).

In all three *Gb* chlorophyll biodegradation products methoxy group of the five membered ring E was not present (Figure 15, 16 and 17).

http://dx.doi.org/10.5772/56155

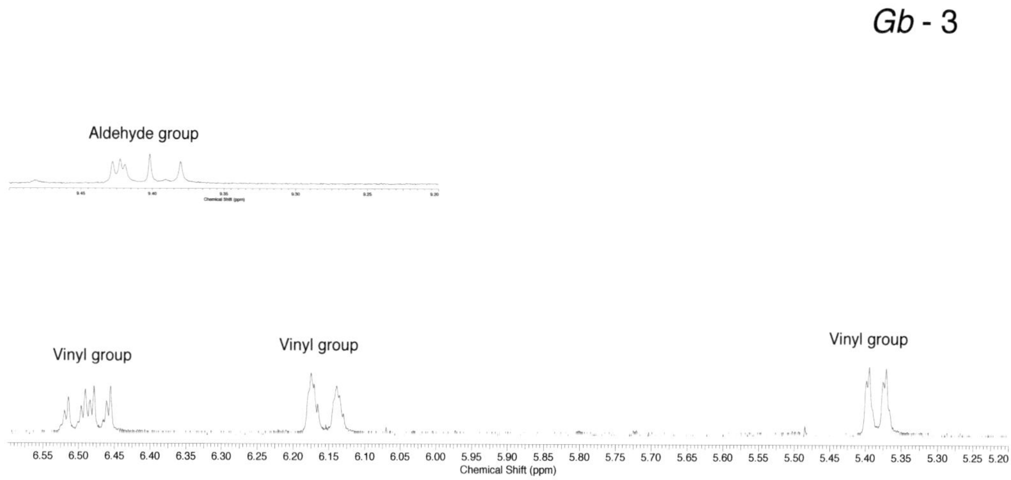

Figure 13. The enlarged proton spectrum of *Gb*-3 in the low field region

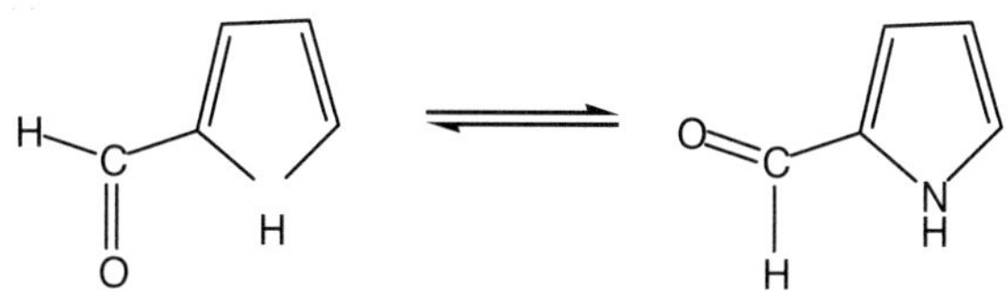

NO-*Z*
14

NO-*E*
15

Figure 14. The conformational change of the pyrrole's aldehyde group

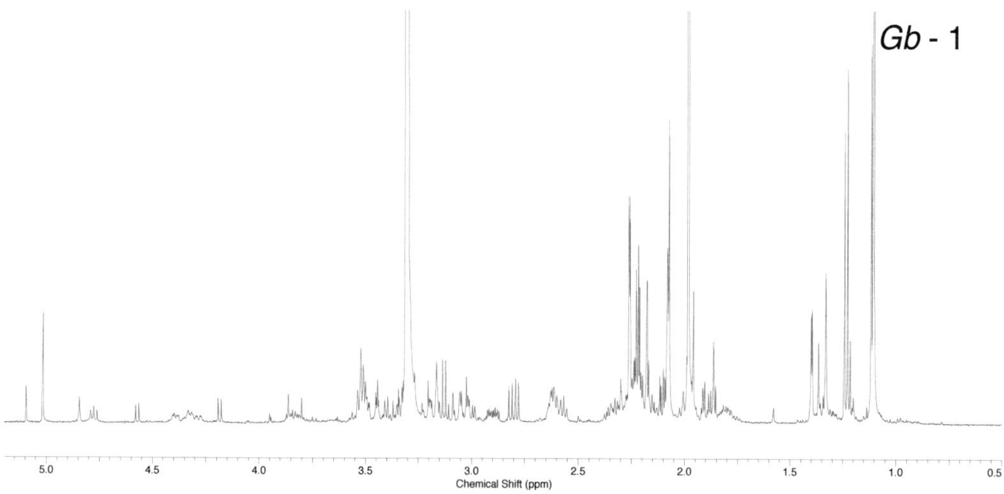

Figure 15. The proton spectrum of the *Gb*-1 in the high field region

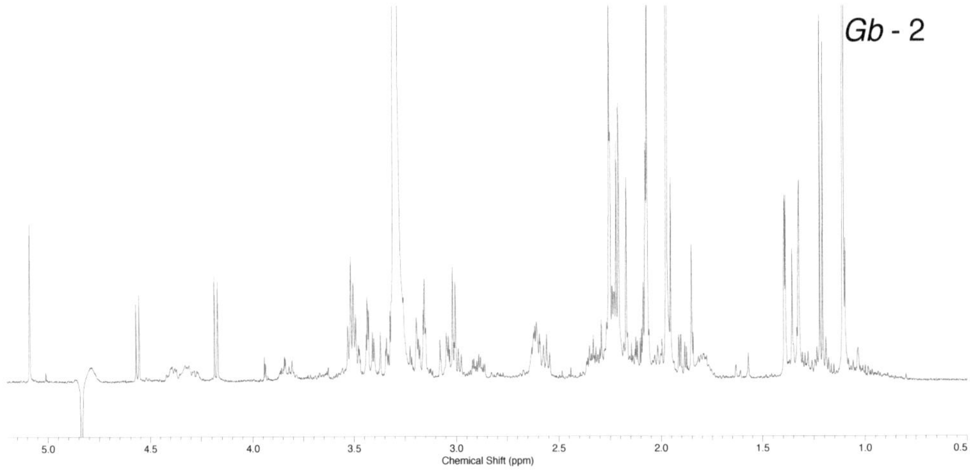

Figure 16. The proton spectrum of the *Gb*-2 in the high field region

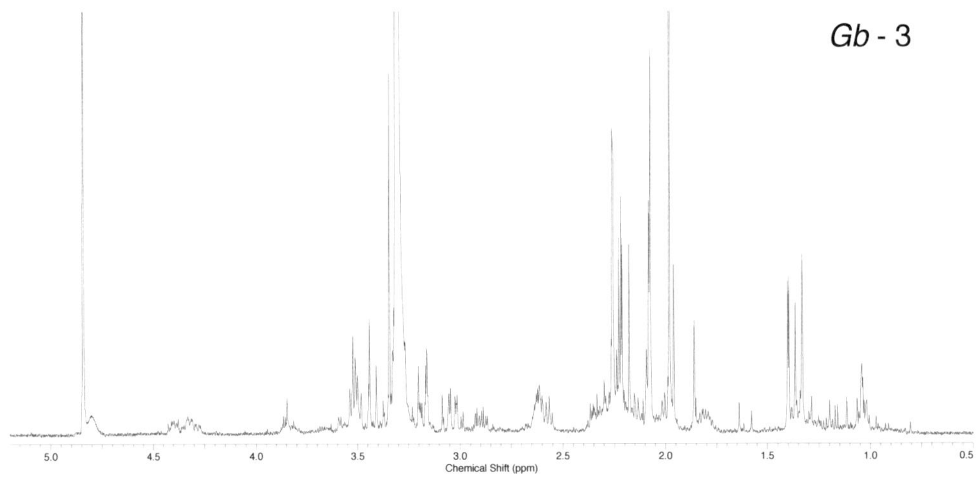

Figure 17. The proton spectrum of the *Gb*-3 in the high field region

The signals of the ethylene protons were found in all spectra recorded by the residual water peak. From the residual water peak to the high field region the presence of one *doublet* and one *doublet of doublets* was present. Those signals are, yet, not assigned. The methyl proton region is not similar to any previous chlorophyll biodegration products' high field region (Figure 15, 16 and 17). The chlorophyll biodegradation products found in *Ginkgo biloba* autumnal leaves are different from all previously isolated ones. The vinyl proton-proton connectivity in the COSY spectrum was easily observed (Figure 18). The *Gb*-3 COSY spectrum recorded was the same as the COSY spectrum of the chlorophyll biodegradation product 9 in the low field region. The establishment of other scalar coupled protons was not determined (Figure 19).

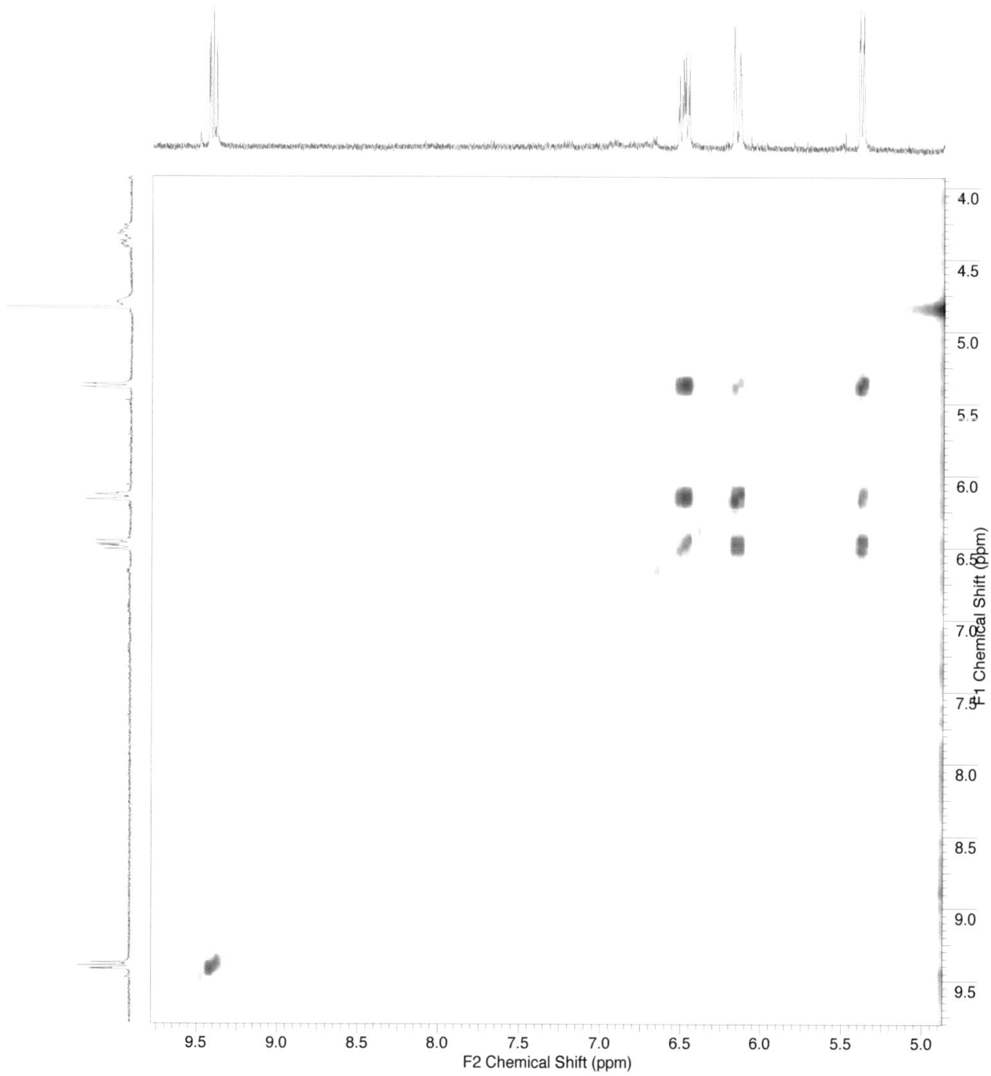

Figure 18. The COSY spectrum of the *Gb*-3 with the spectral expansion in the vinyl region

The dynamic NMR spectra were not recorded, yet, such as, they were done in case of the polycyclic quinones, substituted tetrahydropyrimidines, etc. [28, 29].

The *Ginkgo biloba* chlorophyll biodegradation products were, also, detected in the *Ginkgo biloba* female autumnal leaves' methanol crude extract (Figure 20).

The screening for the presence of the chlorophyll biodegradation products was done in ferns. Of dozens of ferns' autumnal leaves' methanol extracts analyzed only in *Athyrium filix femina*, Dryopteridaceae the *Ginkgo biloba* chlorophyll biodegradation products were detected (Figure 21). The cornifer plant – *Larix kaempferi*, Pinaceae autumnal leaves' methanol crude

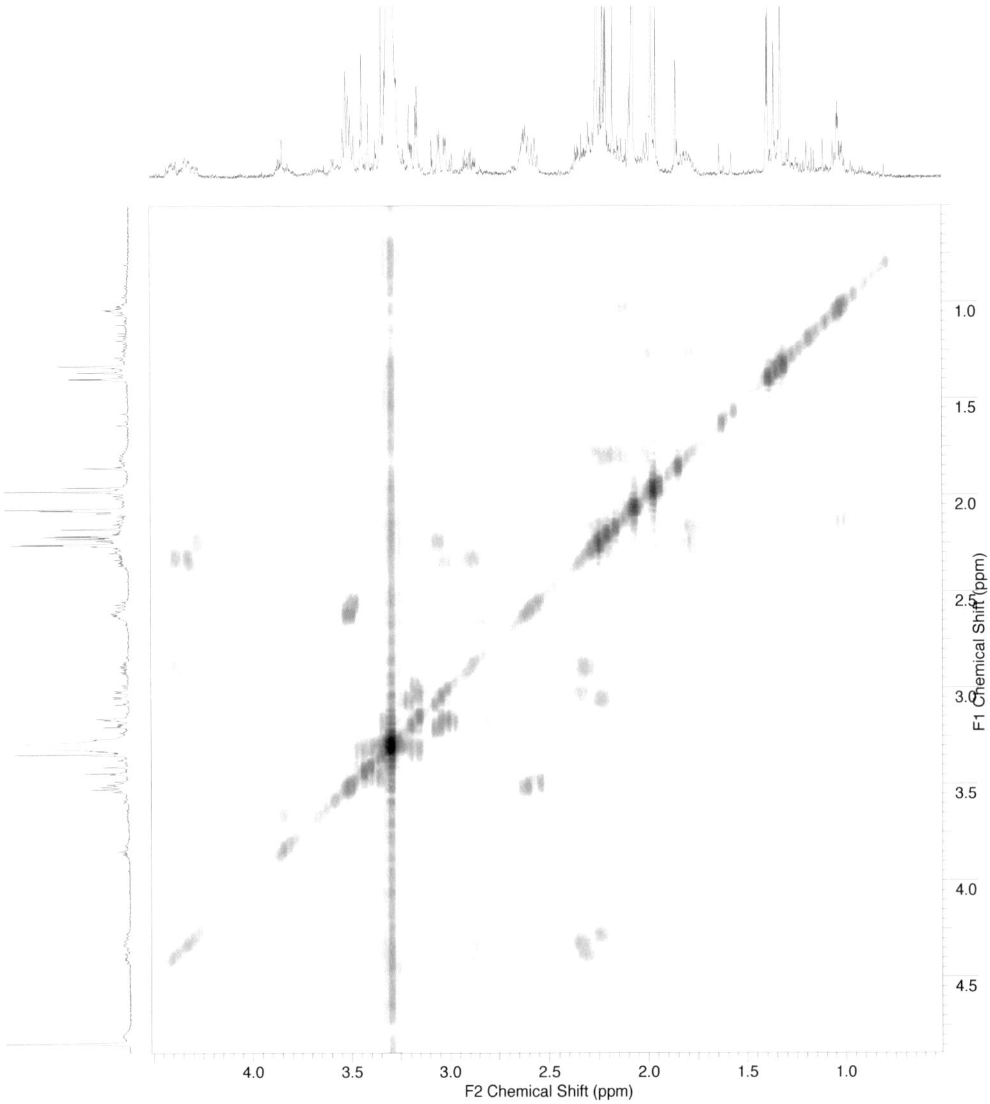

Figure 19. The spectral magnification of the *Gb*-3 COSY spectrum in the high field region

extract was analyzed and the *Ginkgo biloba* chlorophyll biodegradation products' were detected (Figure 21).

The chlorophyll biodegradation product 3 in Hamamelidaceae and Vitaceae autumnal leaves undergoes the oxidation of the aldehyde group at the C5 position forming the urobilinogenic chlorophyll biodegradation product 4.

In the Solanaceae autumnal leaves the chlorophyll biodegradation product 3 undergoes the glycosylation at the $C8^2$ position forming the glycosylated chlorophyll biodegradation product 11.

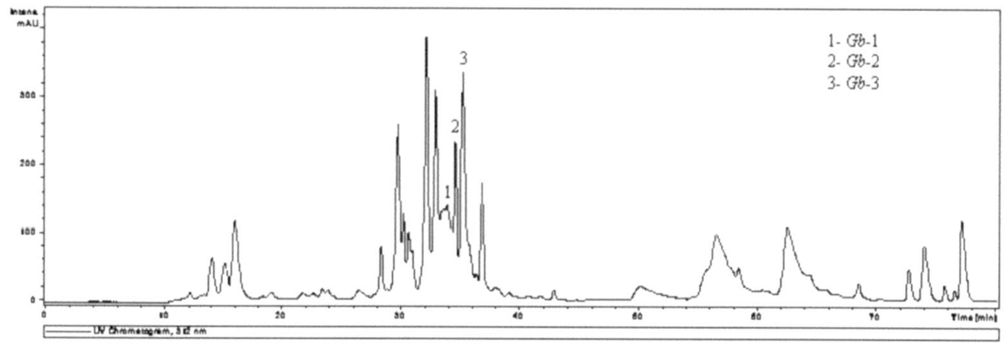

Figure 20. The chromatogram of the *Gingko biloba* female autumnal leaves' methanol crude extract. The LC conditions: Column: Nucleosil 100-5 C_4 4x250 mm. The mobile phase: 90% v/v water (0.1%TFA): methanol to 0% v/v water (0.1%TFA):methanol in 80 minutes. Flow rate: 0.2 ml/min. UV detection at λ=312.

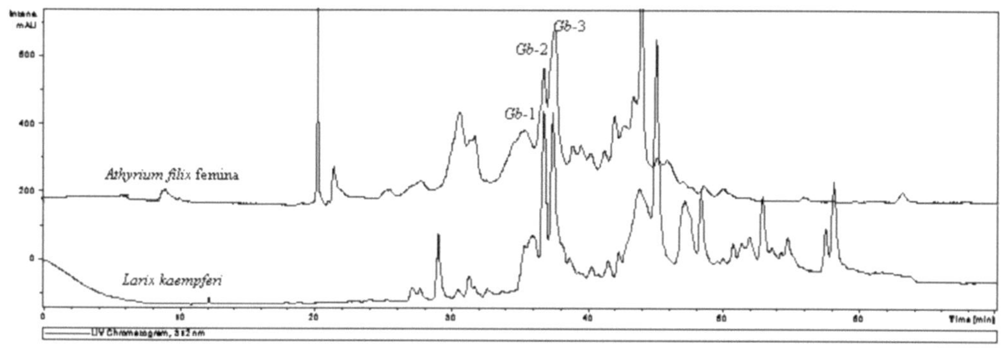

Figure 21. The chromatogram of *Larix kaempferi* and *Athyrium filix femina* autumnal leaves' methanol crude extract. The LC conditions: Column: Nucleosil 100-5 C_4 4x250 mm. The mobile phase: 90% v/v water (TFA): methanol to 0% v/v water (TFA):methanol in 70 minutes. Flow rate: 0.2 ml/min. UV detection at λ=312.

In the *Gingko biloba, Larix kaempferi* and *Athyrium filix femina* autumnal leaves the chlorophyll biodegradation product 3 undergoes the process that still has not been determined.

Further chlorophyll biodegradation products are colourless and were not identified, up to now.

6. The influence of ethylene on chlorophyll degradation

The ancient Egyptians used the ethylene to gas figs in order to stimulate ripening. The ancient Chinese used incense combustion to enhance the ripening of pears. In the 19th century, when the coal gas was used for street illumination, the trees in the vicinity of the street lamps showed extensive defoliation [30]. In 1901, Dimitry Nikolayevich Neljubov, observed that dark-grown pea seedlings growing in the laboratory exhibited symptoms that were later termed the triple response [31]. Dimitry Nikolayevich Neljubov identified ethylene, which was present in the laboratory air from the coal gas, as the molecule causing the response [30]. Dimitry Nikolayevich Neljubov observation can be compared to Leonhard Euler's observation on seven bridges of Koenigsberg, which led to the formation of the Theory of Graphs [32]. Dimitry Nikolayevich Neljubov's observation led to the recognition of ethylene as a plant hormone [30]. In 1910, Herbert Henry Cousins reported that gas evaporation from stored oranges caused the premature ripening of bananas [33]. It was likely that the oranges used were infected with the fungus *Penicillium* which produces large amounts of ethylene. Herbert Henry Cousins showed that plants can produce their own ethylene [33]. In 1917, Doubt concluded that ethylene stimulates abscission [34]. In 1934, Gane reported that plants synthesize ethylene [35]. In 1935, Crocker proposed that ethylene is the plant hormone responsible for fruit ripening [36]. Ethylene, as well, has numerous effects on plant species and organs [30]. The peel of the citrus fruits requires ethylene to induce chlorophyll degradation [37]. The exposure of green leaves to ethylene can induce senescence only in leaves that have reached a defined age [38]. The biochemistry, genetics and physiology of ripening has been extensively studied in economically important fruit crops and the results obtained permitted the formation of the ethylene biosynthesis pathway, signaling and control of gene expression [39]. There is still much to be discovered about these processes at the biochemical and genetic levels.

Author details

Nina Djapic*

Address all correspondence to: djapic@tfzr.uns.ac.rs

Technical faculty "Mihajlo Pupin", University of Novi Sad, Zrenjanin, Serbia

References

[1] Brown, S. B, & Houghton, J. D. Hendry GAF. Chlorophylls. In: Scheer H. (ed.) CRC Press, Boca Raton, FL; (1991).

[2] Robinson, R. Richard Willstätter. (1872). Obituary Notices of Fellows of the Royal Society 1953; , 8(22), 609-634.

[3] Tswett, M. S. O novoy kategorii adsorbtsionnykh yavleny i o primenenii ikh k biokkhimicheskomu analizu. Trudy Varhavskago Obshchestva estevoispytatelei Otd Biol. (1903). , 14-20.

[4] Tswett, M. S. Physikalisch- chemische Studien ueber das Chlorophyll. Die Adsorptionen, Ber. Deutsch. Bot. Ges. (1906). , 24-316.

[5] Tswett, M. S. Adsorptionsanalyse und chromatographische Methode. Anwendung auf die Chemie des Chlorophylls, Ber. Deutsch. Bot. Ges. (1906). , 24-384.

[6] Tswett, M. S. in Chromophylls in Plant and Animal Worlds (1910). Warsaw University, Warsaw (in Russian).

[7] Stoll, A. in Ueber Chlorophyllase und die Chlorophyllide, Dissertation, Eidg. Technische Hochschule, Zuerich, (1912).

[8] Willstaetter, R, & Stoll, A. Untersuchungen uber Chlorophyll. XIX. Ueber die Chlorophyllide, Liebigs Ann. Chem. (1912)., 387-317.

[9] Ružicka, L. Biographical Memoirs of Fellows of the Royal Society (1972)., 18-566.

[10] KrasnovskyJr., AA. Chlorophyll isolation, structure and function: major landmarks of the early history of research in the Russion Empire and the Soviet Union. Photosynthesis Research(2003)., 76-389.

[11] Willows, R. D. Biosynthesis of chlorophylls from protoporphyrin IX, Nat. Prod. Rep. (2003)., 20-327.

[12] Ruediger, W. Biosynthesis of chlorophyll b and the chlorophyll cycle. Photosynthesis Research (2002)., 74-187.

[13] Shioi, Y, Watanabe, K, & Takamiya, K. Enzymatic conversion of Pheophorbide a to the precursor of Pyropheophorbide a in leaves of Chenopodium album, Plant and Cell Physiol. (1996)., 37-1143.

[14] Peoples, M. B, & Dalling, M. J. in Senescence and Aging in Plants (Eds.: L. D. Noodén, A. C. Leopold), Academic press, New York, San Diego, (1988)., 181.

[15] Djapic, N, & Pavlovic, M. Chlorophyll Catabolite from Parrotia Persica Autumnal Leaves. Rev. Chim. (Bucuresti) (2008)., 59(8), 878-882.

[16] Djapic, N, Pavlovic, M, Arsovski, S, & Vujic, G. Chlorophyl Biodegradation Product from Hamamelis virginiana Autumnal Leaves. Rev. Chim. (Bucuresti) (2009)., 60(4), 398-402.

[17] Djapic, N, Djuric, A, & Pavlovic, A. Chlorophyll biodegradation in Vitis vinifera var. Pinot noir autumnal leaves. Research Journal of Agricultural Sciences (2009)., 41(2), 256-260.

[18] Donoghue, N. A, Norris, D. B, & Trudgill, P. W. The purification and properties of cyclohexanone oxygenase from Nocardia globerula CL1 and Acinetobacter NCIB 9871, Eur. J. Biochem. (1976)., 63-175.

[19] Kelly, D. R. A proposal for the origin of stereoselectivity in enzime catalysed Baeyer-Villiger reactions, Tetrahedron: Asymmetry (1996)., 7-1149.

[20] Đapic, N. Behaviour of Fothergilla gardenii chlorophyll catabolite under acidic conditions. Kragujevac J. Sci. (2012)., 34-79.

[21] Djapic, N. private communication. (2012).

[22] Andersen, Ø. M, Aksnes, D. W, Nerdal, W, & Johansen, O-P. Structure elucidation of cyanidin-3-sambubioside and assignments of the ^{1}H and ^{13}C NMR resonances through two-dimensional shift-correlated NMR technique, Phytochemical Analysis, (1991)., 2(4), 175-183.

[23] Pedersen, A. T, Andersen, Ø. M, Aksnes, D. W, & Nerdal, W. Anomeric sugar configuration of anthocyanin O-pyranoside determined from heteronuclear one-bond coupling constants, Phytochemical Analysis (1995)., 6(6), 313-316.

[24] Andersen, Ø. M, Opheim, S, Aksnes, D. W, & Frøystein, N. Å. Structure of petanin, an acylated anthocyanin isolated from Solanum tuberosum using homo- and heteronuclear two-dimensional nuclear magnetic resonance techniques, Phytochemical Analysis (1991)., 2(5), 230-236.

[25] Nerdal, W, Pedersen, A. T, & Andersen, Ø. M. Two-dimensional nuclear Overhauser Enhancement NMR Experiments on Pelargonidin-glucopyranoside, an Anthocyanin of Low Molecular Mass. Acta Chemica Scandinavica, (1992)., 3.

[26] Farnier, M, Drakenberg, T, & Berger, S. Etude, par RMN, des conformations d'aldehydes α et β pyrroliques C-substitues. Stereospecificite des couplages lointains. Tetrahedron Letters, (1973)., 14(6), 429-432.

[27] Farnier, M, & Drakenberg, T. Nuclear Magnetic Resonance Conformational Studies of C-Substituted Pyrrolecarbaldehydes. Part 1. Substituent Effects on Aldehyde Conformations as shown by Long Range Coupling Constants. J. C. S. Perkin II (1975)., 4-333.

[28] Smirnov, A, Fulton, D. B, Andreotti, A, & Petrich, J. W. Exploring ground-state heterogeneity of hypericin and hypocrellin A and B: Dynamic and 2D ROESY NMR study. Journal of the American Chemical Society (1999)., 35(121), 7979-7988.

[29] Garcia, M. B, Grilli, S, Lunazzi, L, Mazzanti, A, & Orelli, L. R. Conformational Studies by Dynamic NMR. 84.[1] Structure, Conformation, and Stereodynamics of the Atropisomers of N-Aryl tetrahydropyrimidines. J. Org. Chem. (2001)., 66-6679.

[30] Taiz, L, & Zeiger, E. Plant physiology. 3rd edition. Sunderland: Sinauer Associates, Inc., MA. (2002).

[31] Neljubow, D. Ueber die horizentale Nutation der Stengel von Pisum sativum und einiger anderen Pflanzen. Beih Bot Zentralbl (1901)., 10-128.

[32] Alexanderson, G. L. About the cover: Euler and Koenigsberg's bridges: A historical view. Bulletin (New Series) of the American Mathematical Society (2006). , 43(4), 567-573.

[33] Cousins, H. H. Agricultural Experiments: Citrus. Annual Report of the Jamaican Department of Agriculture (1910)., 7-15.

[34] Doubt, S. L. The response of plants to illuminating gas. Botan. Gaz. (1917)., 63-209.

[35] Gane, R. Production of ethylene by some fruits. Nature (1934). , 134-1008.

[36] Crocker, W, Hitchcock, A. E, & Zimmerman, P. W. Similarities in the effects of ethylene and the plant auxins. Contributions of the Boyce Thompson Institute (1935). , 7-231.

[37] Purvis, A. C, & Barmore, C. R. Involvement of ethylene in chlorophyll degradation in peel of citrus fuits, Plant Physiol. (1981). , 68-854.

[38] Jing, H-C. Schippers JHM., Hille J., Dijkwel PP. Ethylene-induced leaf senescence depends on age-related changes and OLD genes in Arabidopsis. Journal of Experimental Botany (2005). , 56(421), 2915-2923.

[39] Chaves ALSCelso de Mello-Farias P. Ethylene and fruit ripening: From illumination gas to the control of gene expression, more than a century of discoveries. Genetics and Molecular Biology (2006). , 29(3), 508-515.

Chapter 3

Bioremediation of Agricultural Land Damaged by Tsunami

M. Azizul Moqsud and K. Omine

Additional information is available at the end of the chapter

http://dx.doi.org/10.5772/56595

1. Introduction

Bioremediation is the process of using naturally occurring microbes to digest and convert unwanted material into harmless substances. In this chapter, an innovative study has been carried out by using compost and specific bacteria (Halo Bacteria) to restore the high saline soil damaged by Tsunami occurred on 11th March, 2011 at the Tohoku Area in Japan.

A disaster is the tragedy of a natural or human-made hazard (a hazard is a situation which poses a level of threat to life, health, property, or environment) that negatively affects society or environment. A natural disaster is a consequence when a natural hazard (e.g., volcanic eruption or earthquake) affects humans. Tsunamis and earthquakes are two of the most dangerous and yet most common hazards to affect population centers and economic infrastructures worldwide. Generally, tsunami flooding results from a train of long-period waves that can rapidly travel long distances from where they were generated by deep-ocean earthquakes, submarine landslides, volcanic eruptions, or asteroid impacts [1, 2, 3]. Due to tsunami the sea water carry sediments along with salt itself. There have been many studies on recent and ancient tsunami deposits. These include descriptions of tsunami deposits in coastal lake, estuary, lagoon, bay floor and shelf environments and even the farmland [4,5]. The mega earthquake and consequent tsunami had caused a great damage to not only human life and infrastructure but also the agricultural land and the crops in Tohoku region, Japan. The after math of the tsunami has created many problems to environment and geo-environment of these affected areas. Soil pollution and high salinity which caused the farmland unusable for cultivation is one of the major geo-environmental problems. The objective of this study is to get an idea about the extent of soil chemical properties change due to tsunami and to apply bioremediation approach to salinity control of the agricultural land.

The sea water inundated the large areas of agricultural land causing the excessive saline in the soil.

2. Salinity of soil

Salinity of soils is the condition of soils that have a high salt content. The predominant salt is normally sodium chloride (NaCl). As a result, saline soils are therefore also *sodic soils* but there may be sodic soils those are not saline, but alkaline. Salty soils are a common feature and an environmental problem in irrigated lands in arid and semi-arid regions all over the world. They have poor or little crop production. The problems are often associated with high water tables, caused by a lack of natural subsurface drainage to the underground. Poor subsurface drainage may be caused by insufficient transport capacity of the aquifer or because water cannot exit the aquifer. Worldwide, the major factor in the development of saline soils is a lack of precipitation. Most naturally saline soils are found in (semi)arid regions and climates of the globe.

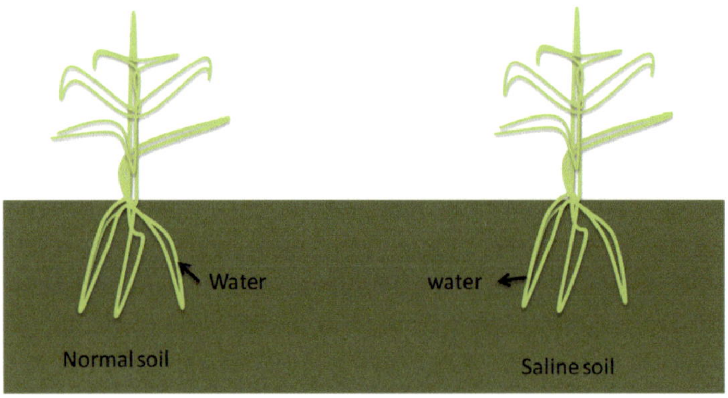

Figure 1. Mechanism of salinity affected soil and plant interaction

Figure 1 shows the mechanisms of salinity affected soil and plant interaction. Salinity becomes a problem when enough salts accumulate in the root zone to negatively affect plant growth. Excess salts in the root zone hinder plant roots from withdrawing water from surrounding soil. This lowers the amount of water available to the plant, regardless of the amount of water actually in the root zone. For example, when plant growth is compared in two identical soils with the same moisture levels, one soil receiving salty water and the other receiving salt-free water, plants are able to use more water from the soil receiving salt-free water. Although the water is not held tighter to the soil in saline environments, the presence of salt in the water causes plants to exert more energy extracting water from the soil. The main point is that excess salinity in soil water can decrease plant available water and cause plant stress. So, high salinity of soil is very dangerous for the plant as most of the plants can not survive in that soil condition.

http://dx.doi.org/10.5772/56595

Figure 2. Salt accumulated on the surface of the tsunami sediment in Tohoku area, Japan

Figure 2 shows the salt accumulation on the surface of the soil after the tsunami in Tohuku area in Japan. For the case of tsunami, a vast area of the land area goes under sea water. And the accumulation of salt after the tsunami water caused a serious damage to the geo-environment. Soil water salinity can affect soil physical properties by causing fine particles to bind together into aggregates. This process is known as flocculation and is beneficial in terms of soil aeration, root penetration, and root growth. Although increasing soil solution salinity has a positive effect on soil aggregation and stabilization, at high levels salinity can have negative and potentially lethal effects on plants. As a result, salinity cannot be increased to maintain soil structure without considering potential impacts on plant health.

3. Tohoku region Pacific Coast earthquake in 2011

The great east Japan Earthquake (Higashi Nihon Daishinsai in Japanese) was a magnitude 9.0 undersea mega thrust earthquake off the coast of Japan that occurred at 14:46:23 JST on Friday, 11 March 2011.The location of the epicenter (38.322^0 N, 142.369^0 E) of this earthquake is about 70 kilometers east of the Oshika Peninsula of Tohoku and the hypocenter at an underwater depth of approximately 32 km. It was the most powerful known earthquake to have hit Japan, and one of the five most powerful earthquakes in the world overall since modern record-keeping began in 1900. The earthquake triggered extremely destructive tsunami waves of around 40 m in Miyako, Iwate prefecture and in some cases travelling up to 10 km inland. In addition to loss of life and destruction of infrastructure, the tsunami caused a number of nuclear accidents in the power plant in Fukushima which caused evacuation zones affecting hundreds of thousands of residents. The sea water inundated the large areas of agricultural land causing the excessive saline in the soil.

4. Soil investigation

The field test of soil for its chemical analysis was conducted in Rikuzentakata city of Iwate prefecture, Japan. Fig. 3 shows the damaged area in Rikuzentakata city [9]. This city was one the major affected areas by the tsunami on 11th March, 2011. Fig. 4 shows the place of investigation on 5th May, 2011. Fig. 5 shows the place of soil investigation on 30th June, 2011.

The pH and EC (Electric Conductivity) of the damaged agricultural land were measured by the digital pH meter (Horiba, D-54SE). The electrical conductivity of the soil was also measured by using digital EC meter (Oakton, PCSTEST35). The salinity of the soil can be calculated by the value of electrical conductivity.

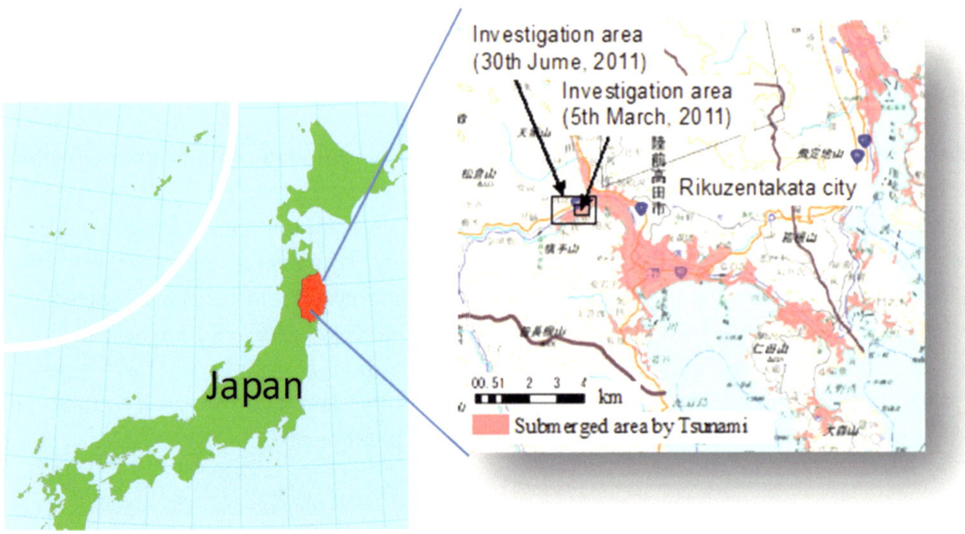

Figure 3. Map of Japan indicating the study area and soil investigation area

5. Soil properties of Rikuzentaka area after tsunami

The soil of the Rikuzentakata was very fertile. The farmers used to cultivate different types of crops including the corn and vegetables. The average moisture content of the soil was around 25 %. However, due to the tsunami water inundated the large areas of the Rikuzentakata, a huge amount of submarine sediments come along with the sea water and settled on the agricultural land along with different types of tsunami debris. The sediment of the tsunami also brought some kinds of toxic materials which were settled under the sea over a long period of time. The sediment is mainly a clay but some sandy particles were also found in different parts of the area.

http://dx.doi.org/10.5772/56595

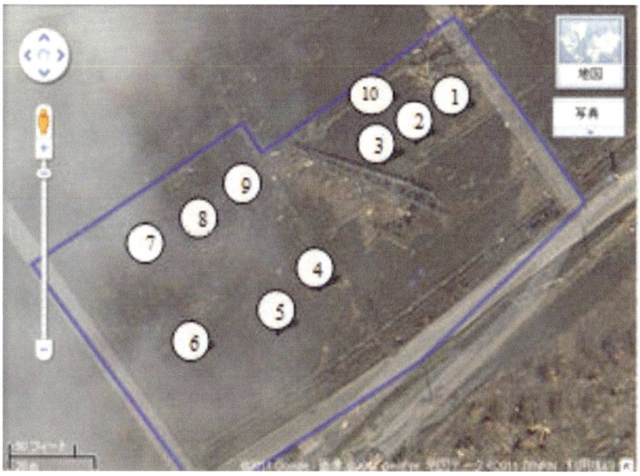

Figure 4. The place of soil investigation on 5 May 2011 in Rikuzentakata area

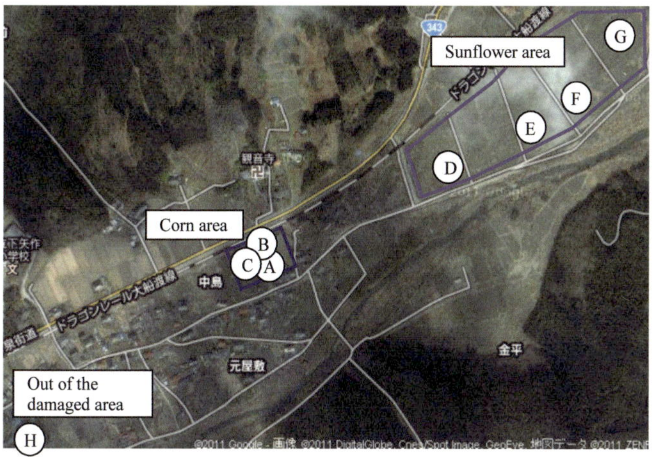

Figure 5. The place of soil investigation on 30th June 2011 in Rikuzentakata area

6. Bioremediation for restoration of saline soil

The aim of soil salinity control is to prevent soil degradation by salinization and reclaim already salty (saline) soils.

Various attempts are now carrying out to control the salinity of the agricultural land. The primary method of controlling soil salinity is to permit 10-20% of the irrigation water to leach the soil, be drained and discharged through an appropriate drainage system.

The salt concentration of the drainage water is normally 5 to 10 times higher than that of the irrigation water, thus salt export matches salt import and it will not accumulate. However, it

Figure 6. Tsunami Sediment deposited on the vast area in tohoku, Japan

will take a long time and efforts for such kind of design of the salinity removal from the saline soil [7,8].

In this study, an innovative idea has been taken for reducing the salt concentration from the soil of the agricultural areas by bioremediation. By using the halo bacteria with the compost the bioremediation was carried out.

It was possible to increase a volume of the compost by mixing rice bran, oil cakes, grinds of fish bones and water in a specific ratio. Total 30 kg compost was made. Further, it had increased up to 300 kg. After mixing each material, temperature of the compost increased at 48 °C for 2 days and turned over for aeration (Figs. 7 and 8).

Figure 7. Preparing compost by mixing rice bran, oil cakes, grinds of fish bones and water

http://dx.doi.org/10.5772/56595

Figure 8. Compost by mixing rice bran, oil cakes, grinds of fish bones and water after 3 days

Then, the compost containing the halo bacteria had been applied in the large areas of Rikuzentakata of Iwate Prefecture which was affected by sea water inundation due to tsunami. Due to bacterial activities, the salinity of the agricultural areas would have been reduced as well as compost would supply some nutrients and organics to the soil.

(a) EC (mS/cm)

Depth	1	2	3	4	5	6	7	8	9	10
5cm	1.51	0.36	0.39	1.49	1.26	0.47	1.20	1.77	1.01	0.25
10cm	3.04	2.00	1.94	2.93	3.16	1.32	2.77	3.43	1.83	1.15
15cm	2.23	2.43	2.81	4.03	2.21	1.84	0.97	3.46	2.4	2.25

(b) pH

Depth	1	2	3	4	5	6	7	8	9	10
5cm	7.21	8.47	8.06	7.55	7.22	8.39	7.71	7.62	7.19	8.27
10cm	5.75	5.68	6.5	5.78	5.71	6.57	5.98	5.6	6.74	6.52
15cm	6.22	5.48	5.69	5.2	6.13	5.84	7.12	5.64	5.77	5.82

Table 1. EC and pH of soil in different depths on 5th May 2011

(a) EC (mS/cm)

Depth	A	B	C	D	E	F	G	H
Surface	0.56	0.11	0.13	0.083	0.09	0.064	0.13	-
5 cm	-	-	-	-	-	-	-	0.032
10 cm	0.97	1.19	0.62	-	0.093	0.15	-	-
15 cm	-	-	-	-	-	-	0.21	-
20 cm	1.83	1.58	1.64	0.21	0.096	0.46		0.026
25 cm	2.15	-	-	-	-	-	-	-
30 cm	-	-	1.65	-	-	-	-	-

(b) pH

Depth	A	B	C	D	E	F	G	H
Surface	7.2	8.1	7.4	8.3	9.0	8.4	8.0	-
5 cm	-	-	-	-	-	-	-	6.9
10 cm	6.6	5.9	6.7	-	9.1	8.4	-	-
15 cm	-	-	-	-	-	-	6.7	-
20 cm	6.1	5.6	5.9	7.0	8.0	7.1	-	6.9
25 cm	6.4	-	-	-	-	-	-	-
30 cm	-	-	5.7	-	-	-	-	-

Table 2. EC and pH of soil in different depths on 30th June, 2011

7. Laboratory test of Bioremediation with the compost

In order to confirm the effect of the compost, an artificial saline soil was made by mixing natural salt. The original EC of the soil is 2.85 mS/cm. The compost with Halo bacteria of 1g and rice bran of 15 g were mixed into the soil of 500 cm^3. The rice bran is a material for accelerate growth of the bacteria. After the incubation, a value of EC is measured. Fig. 9 shows the result of EC for the soil mixed with the compost. It is clear that EC of the soil with salt is decreased by mixing the compost with Halo bacteria. It is estimated that 25% of salt can be reduced by mixing the compost for 7 days in this bioremediation.

The bioremediation of the saline soil is possible by using the home made compost. The way of making the compost is easy and less costly. So, this method of remediation can be used in many developing countries in the world. The benefit of this bioremediation is that the cost during the process of bioremediation is very low compare to other method of salinity management of agricultural soil currently used.

Soil investigation at the site was performed on March 2012. Table 3 shows comparison of electric conductivity of the soils before and after the restoration by the compost. Due to the rail fall and vegetation of sunflower, the salt concentration decreased gradually and the highest EC at the site was 0.25 mS/cm on September 2011. The value of EC decreased furthermore on

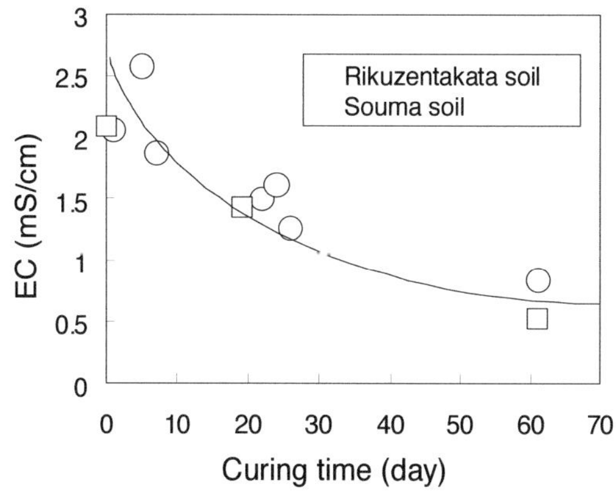

Figure 9. Relationship between electric conductivity and curing period on soil with salt

March 2012.The effect of bioremediation is understood in the field, however, the exact amount of the bioremediation is difficult to measure in the field as the natural environmental effects have the influence on the soil properties in the wide areas of the Rikuzentakata.

			EC (mS/cm)				
Date	A	B	C	D	E	F	G
30/9/2011	0.174	0.213	0.211	0.249	0.186	0.203	0.226
12/3/2012	0.016	0.017	0.018	0.143	0.017	0.016	0.017

Table 3. Comparison of electric conductivity of the soils before and after bioremediation

Some substances known to have toxic properties have been introduced into the environment through man-made activities. These substances range in degree of toxicity and danger to human health. Many of these substances either immediately or ultimately come in contact with or are sequestered by soil. Conventional methods to remove, reduce, or mitigate toxic substances introduced into soil or ground water via anthropogenic activities and processes include pump and treat systems, soil vapor extraction, incineration, and containment. Utility of each of these conventional methods of treatment of contaminated soil and/or water suffers from recognizable drawbacks and may involve some level of risk.

8. Mechanisms of bioremediation of salinity affected soil

The mechanism of bioremediation of the salt affected agricultural land is illustrated in the Figure 10. It is seen that the bacteria mixing with compost help to release the salt element from the soil surface due to the microbial activities. When the effects of bioremediation were taken place then the salt removed easily from the surface with the help of rain water or snow melting.

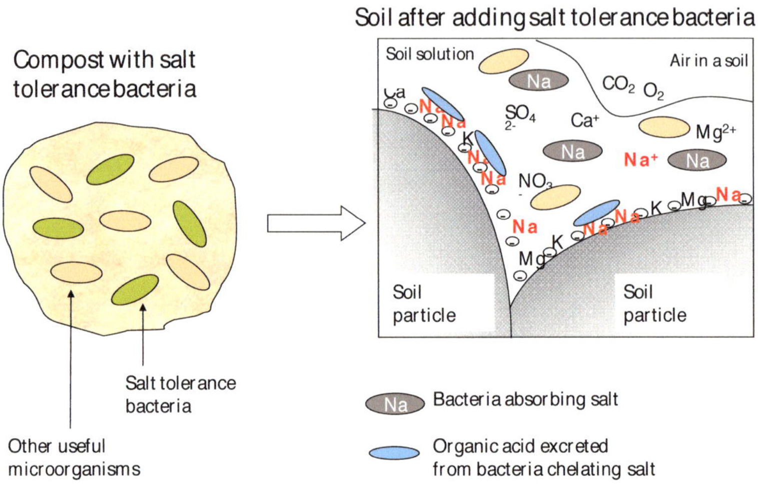

Figure 10. Mechanism of salt removal from tsunami affected soil by bioremediation

The salt in the soil particle can be easily removed by this bioremediation. So this mechanism to remove the salt from the saline soil can be used not only for the tsunami affected areas but also in the coastal areas in different countries in where the sea level rise caused by global warming is a real threatening matter for the local population.

9. Salinity, plant growth and bioremediation in tsunami affected area

Salinity is an important land degradation problem. Soil salinity can be reduced by leaching soluble salts out of soil with excess irrigation water. Soil salinity control involves watertable control and flushing in combination with tile drainage or another form of subsurface drainage however, if the vast area is affected by salinity then it is really difficult to treat that soil as the fresh waster needed to wash the soil will need a huge amount of money. So bioremediation of saline soil is a better option as the compost can supply some nutrients as well to the soil. High levels of soil salinity can be tolerated if salt-tolerant plants are grown. Sensitive crops lose their vigor already in slightly saline soils, most crops are negatively affected by (moderately) saline soils, and only salinity resistant crops can stay alive in severely saline soils.

A high salt level interferes with the germination of new seeds. Salinity acts like drought on plants, preventing roots from performing their osmotic activity where water and nutrients move from an area of low concentration into an area of high concentration. Therefore, because of the salt levels in the soil, water and nutrients cannot move into the plant roots.

As soil salinity levels increase, the stress on germinating seedlings also increases. Perennial plants seem to handle salinity better than annual plants. In some cases, salinity also has a toxic effect on plants because of the high concentration of certain salts in the soil. Salinity prevents the plants from uptaking the proper balance of nutrients they require for healthy growth. So, in our bioremediation of saline soil, we can easily provide the sufficient nutrients as well as to restore the salinity affected soil for a vast area generally affected by tsunami in a reasonable cost.

10. Conclusions

The mega earthquake and consequent huge tsunami has done a great damage to the entire areas of the pacific regions in Tohoku, Japan. The sea water which covered the agricultural lands in these areas has created a critical situation for the farmers. The farmers have lost not only the crops they were cultivating but also the soil of the agricultural field had been seriously damaged by the sea water, salinity and other pollutants. The pH value and EC value of the soil in these areas are considered as the higher value in terms of safer limit for the regular crops. To reclaim this saline soil, compost containing the Halo bacteria had been applied as an approach of bioremediation. The Halo bacteria used the excessive salts from the soil and consequently can reduce the salinity problem which was proved in the laboratory test. This compost can also provide necessary nutrients to the soil and plant. So, bioremediation by compost to restore the tsunami damaged saline soil proved to be an efficient and can be applicable in other parts of the world especially developing countries which are suffering by the sea level rise problems.

Author details

M. Azizul Moqsud[1] and K. Omine[2]

1 Department of Civil Engineering, Yamaguchi University, Japan

2 Department of Civil Engineering, Nagasaki University, Japan

References

[1] Morton, R.A., G. Gelfenbaum and B.E. Jaffe (2007) Physical criteria for distinguishing sandy tsunami and storm deposits using modern examples. *Sedimentary Geology*, vol. 200,,pp.184-207.

[2] Katsudanso Kenkyu (Active Fault Res. Jpn.)12, pp 1-23 (in Japanese with English abstract).

[3] Minoura K.,Gusiakov, V.G.,Kurbatov,A., Takeuti,S.,Svendsen, J.I., Bondevik, S.,Od, T.(1996). Tsunami sedimentation associated with the 1923 kamchatka earthquake. *Sedimentary Geology*, Vol.106, pp.145-154.

[4] Bourgeois J., T.A. Hansen, P.L. Kauffman,E.G. (1988). A tsunami deposit at the Cretaceous-Tertiary boundary in Texas. Science, Vol. 241, pp. 567-570.

[5] Dawson A.G., Long, D., Smith, D.E., (1988). The steregga slides; evidence from eastern Scotland for possible tsunamis. Marine Geology. Vol.99,pp.265-287.

[6] Yamazaki T., Yamaoka,M.Shiki,T.(1989). Miocene offshore tractive current-worked conlomerates-Tsubutegaura, Chiba Peninsula, Central Japan. In: Taira,A.,Masuda,F(Eds). Sedimentary Facies in the Active Plate Margin, TERRAPUB, Tokyo, pp. 483-494.

[7] Shiki T and Yamazaki,T.,(1996). Tsunami-induced conglomerates in Miocene upper bathyal deposits, Chiba Peninsula, Central Japan. *Sedimentary Geology*. Vol 104, pp. 175-188.

[8] Wright C.,Mella, A. (1963). Modifications to the soil pattern of south-central Chile resulting from seismic and associated phenomena during the period May to August 1960. Bull.Seismol.Soc.Am.Vol 53,pp. 1367-1402.

[9] The Geospatial Information Authority of Japan.

Chapter 4

Microbial Degradation of Some Halogenated Compounds: Biochemical and Molecular Features

Yu-Huei Peng and Yang-hsin Shih

Additional information is available at the end of the chapter

http://dx.doi.org/10.5772/56306

1. Introduction

Due to the advance of organic and synthetic chemistry and many applications of man-made organic compounds, lots of xenobiotic chemicals are produced and benefit our life. However, some of them are persistent in the environment and their toxicities are accumulated through the food webs. Due to the potential hazard for human and the ecosystem, the regulations on the usage of these persistent organic pollutants (POPs) and the development of safe decomposition methods are now in great request.

Tetrachloroethene (PCE) and trichloroethene (TCE) (Table 1) have been widely used as dry cleaning solvents and degreasing agents. Due to poor disposal practices and accidental release, they are within the most-abundant groundwater contaminants. Exposure to PCE is injurious to epidermis, kidney and nervous system [1]. It has been classified as a probably carcinogen [2]. Exposure to TCE leads to acute effects on liver, kidney, central nervous, and endocrine systems. It is also associated with several types of cancers based upon epidemiological research [3]. PCE and TCE are regulated in U.S.A. to a maximum contaminant level of 5 ppb. The use of PCE and TCE in the food and pharmaceutical industries has been banned across much of the world since the 1970s. However, these chemicals are still used as a degreasing agent for other demand. Besides, the mono-chlorinated ethene, vinyl chloride (VC), is known as carcinogen that causes liver cancer [4] and the *cis*-1,2-dichloroethene (cDCE) is harmful to nervous system, liver and blood cells [5].

Polybrominated diphenyl ethers (PBDEs), composed by two phenol rings and linked by one oxygen atom (Table 1), allow maximum ten bromide atom incorporated on the phenol rings to form 209 possible congeners. They have been widely used as flame retardants in many products over more than three decades. Their usage has protected both human lives and their properties from fire damage. PBDEs disrupt the balance of thyroid hormone, lead to reproductive toxicity, hepatic toxicity, immunotoxicity and developmental neurotoxicity in mammals [6, 7]. The toxicity of PBDEs and their metabolites are due to elevated free radicals, DNA damages, cell cycle blockage and apoptosis rate [8, 9]. Among the congeners of PBDEs, the usage of penta-BDEs and octa-BDEs has been banned in the European Union and several states of the USA; tetra-BDE to hepta-BDE have also been classified as POPs and the production of decabromodiphenyl ether (DBDE) will cease in 2013 in USA. However, the concentration of PBDEs in environment remains exponentially increasing because of the consequence of long-term usage [10]. Due to their ubiquitous distribution in the environment, potential toxicity, tendency for bioaccumulation, and the increased accumulation amount in the environment, the fates of PBDEs in the nature is serious concern for public health.

Hexabromocyclododecane (HBCD), a brominated aliphatic cyclic hydrocarbon (Table 1), is another widely used brominated flame retardant (BFR). It has animal thyroidal and developmental toxicity. The toxicity is due to altering the expression and function of metabolic enzymes, increasing hormone turn over and apoptosis [11, 12]. HBCD has been detected widely in biota and abiotic samples [11]. Due to its persistent, bioaccumulative, and toxic properties, HBCD has been proposed as a Substance of Very High Concern under the REACH regulations [13] and included on the USEPA's lists of Chemicals of Concern [14]; It is also under screening-level risk assessments to determine if it meets criteria of compounds in the Stockholm Convention and in the UN-ECE Protocol on POPs [15].

Most traditional remediation methods are not suitable for degrading chloroethenes, PBDEs and HBCD. For example, dehalogenation processes under oxidative, alkaline, or irradiation conditions are high cost of energy and treatment reagents [16, 17]. Pyrolysis is only limited for specific contaminated media with high heat conductance [18]. The generation of hazardous by-products is also a problem [19] It has been reported that PBDEs and HBCD could be photodegraded [20]; however, pollutants accumulated in the soils, sediments, water bodies are not easy approach to light. Recently, the permeable reactive barrier made by zerovalent iron offers a new direction for halogenated compounds remediation [21]. Electrons offered from iron can reduce the halogenated compounds through reductive dehalogenation.

Zerovalent iron is cheaper than above processing methods, and the shortage of low efficiency is compensated by newly developed nanotechnology. Nanoscale zerovalent metals can degrade chloroethenes, PBDEs, and other contaminants with a fast kinetics and high efficiency [22-26]. There are still some limitations by using nano-metals, such as: toxic by-products generation due to incomplete dehalogenation [23], potential hazardous effect from nanoparticles [27], and large requirement for metals. Therefore, it comes to be one of the recent trends in developing nanomaterials with high efficiency and low environmental impact, and combining with other treatment technologies.

Microbiological approaches produce less intervention to the environment and are less expensive than physical or chemical methods. Biodegradation of chloroethenes has been extensively investigated and reviewed [28-30]. Bioremediation for PBDEs and HBCD are just at the beginning. The main objective of this review is to summarize current knowledge of microbial degradation of chloroethene, PBDEs and HBCD, especially from the biochemical and molecular point of view. We also attempt to compare the advantages and drawbacks of the combined approaches which may apply to field remediation.

2. Biodegradation of chloroethenes, PBDEs and HBCD

Biodegradation of chloroethenes, PBDEs and HBCD occurs in various environmental or living samples [31, 32]. In the environment, microorganisms play major roles in the degradation reactions; while intrinsic detoxicification systems in plants and animals bodies metabolize these compounds [31, 33]. In this article, the diverse and complex microbial degradation machineries are presented and compared.

Biotransformation of chloroethenes, PBDEs and HBCD in aerobic or anaerobic environments has been demonstrated (Fig. 1). In aerobic environment, chloroethenes and PBDEs are metabolized with the generation of energy or degraded cometabolically without energy-yield. In anaerobic condition, they are reduced through the energy yielded from the oxidation of electron donors, i.e. reductive dehalogenation or dehalorespiration. Biotransformation of HBCD might mediate through hydrolytic dehalogenation, which may occur either in aerobic or anaerobic conditions. Detail for the each type of reaction and the degraders will be described in the following sections.

2.1. Aerobic oxidative degradation

Under aerobic conditions, chloroethenes and PBDEs can be oxidized both cometabolically and metabolically (Fig. 1, left part). Metabolic degradation indicates the use of the above compounds as growth substrate. Chloroethenes and PBDEs more easily undergo aerobic transformation with less numbers of halogen substituent.

Metabolic degradation of cDCE and VC as sole carbon and energy source has been reported by many bacteria, such as the *Pseudomonas* sp. and *Bacillus* sp. [34, 35]. Using cDCE as auxiliary substrate for growth is much less [35] and not shown in TCE and PCE. After oxidative transformation, the auxiliary substrate may be mineralized or the carbon atoms may be incorporated into biomass. Microbial growth can be confirmed by monitoring the stable isotope fractionation and is suitable for field assessment [36].

On the contrary, cometabolic degradation of chloroethenes occurs fortuitously during the degradation of growth dependent substrates (auxiliary substrates), such as methane, ammonia, or aromatic hydrocarbons. Even cDCE can be cometabolized when VC is metabolic degraded [34]. Cometabolic degradation of TCE, DCE and VC is common [37, 38]. So far *Pseudomonas stutzeri* OX1 is the only one that could aerobically cometabolize PCE [39]. Therefore, without primary substrate supplement, intrinsic bioremediation with air or nutrients injection alone could not enhance the aerobic cometabolic mechanism and would not cause the microbial degradation of PCE and TCE contaminated sites.

PBDEs could be degraded into phenol or catechols by aerobic microbial through hydroxylation or bond cleavage [33] (Table 2). *Sphingomonas* sp. SS3 and SS33 can transform mono- or dihalogenated DEs for growth [40, 41]. In addition, *Sphingomonas* sp. PH-07 could break down several lower-bronimated BDE congeners (up to tri-BDEs) [42]. Other PBDEs degradation bacteria are reported [43]: *Rhodococcus jostii* RHA1 and *Burkholderia xenovorans* LB400 transform several lower-brominated BDE congeners (up to penta- and hexa-BDEs); *Rhodococcus* sp. RR1 transforms di- and mono-BDEs and the *Pseudonocardia dioxanivorans* CB1190 only degrades mono-BDEs. The transformation by-products include phenol, catechol, halophenol and halocatehol, indicating nonspecific attractions. These degraders might transform PBDEs through cometabolic reactions because auxiliary substrates such as diphenyl ether are supplemented. The *Lysinibacillus fusiformis* strain DB-1, cometabolically debrominate DBDE with the metabolism of lactate, pyruvate and acetate, is isolated [44].

So far, there is only one degrader been reported can transform HBCD: *Pseudomonas* sp. HB01 [45]. Since bromide atom was not detected after degradation reaction, such transformation might not through haloelimination. In general, cometabolism requires supplement of auxiliary substrates and there is no energy yielded. Therefore, microorganisms do not favor proceed this kind of reaction. Besides, the dehalogenation reaction is usually incomplete, resulting accumulation of toxic intermediates. The contribution on the bioremediation from cometabolism is limited [30].

2.2. Anaerobic reductive dehalogenation

Reductive dehalogenation is an anaerobic respiration process. Electron donors are oxidized and the halogenated compounds are reduced through accepting the electrons. The free energy generated from this reaction supports the growth of microbial degraders. Hydrogen atom replaces the halogen atoms one after another resulting in the dehalogenation sequence from higher-numbered compounds to lower-numbered ones. Contrary to aerobic degradation, the potential for reductive dehalogenation increases with the number of halogenated substituent [29].

Hydrogen gas is generated primary by fermentative and acetogenic bacteria (Fig.1, right part). The dehalorespiration bacteria compete with hydrogenotrophs, such as sulphate-reducers, nitrate-reducers, methanogens, acetogens, and other reducers [46, 47]. Except hydrogen, other electron donors also can be used for reductive dehalogenation, ex. *Sulfurospirillum multivorans* can also use pyruvate and formate [48].

Several mixed cultures and pure strains are known to reductively transform chloroethenes. Mixed culture could cooperate and transform PCE to ethene. The pure strains belong to different genus, such as Bacillus, Dehalobacter, Dehalococcoides, Desulfitobacterium, Geobacter, and Sulfurospirillum (Table 2). Most of them only dechlorinate PCE and TCE to cDCE. Only *Dehalococcoides ethenogenes* strain 195 can reductively dechlorinate PCE to ethene. The accumulated hazardous DCE and VC is a major obstacle in bioremediation of chloroethene contaminated sites. *Dehalococcoides* sp. strain BAV1, dechlorinates DCE and VC and cometabolizes PCE and TCE; the accumulated toxic compounds can be transformed into benign ethene [49].

Reductive debromination of PBDEs has been reported through pure strains (Table 2) or mixed cultures. Most of the debromination processes require TCE to be co-substrate. 20 mixed microcosms can degrade octa-BDE mixture to hexa- to mono-BDEs within 2 months [50]. *Sulfurospirillum multivorans* could debrominate DBDE into hepta- and octa-BDEs after 2 months of incubation. *D. ethenogenes* strain 195 could debrominate the octa-BDE mixtures into hepta- to di-BDEs after 6 months of incubation [51]. *Dehalococcoides* sp. Strain DG could degrade octa-BDE mixture into terta- and penta-BDEs or transform penta-BDE mixture into terta-BDE [52]. Several dechlorinating bacteria, *Desulfitobacterium hafniense* PCP-1, *Dehalobacter restrictus* PER-K23, *Desulfitobacterium chlororespirans* Co23 and *Desulfitobacterium dehalogenans* JW/IU-DCI debrominate the octa-BDE mixture and the most frequently detected congeners, penta 99 and tetra 47 when PCP, PCE, 3-chloro-4-hydroxybenzoate, or 3-chloro-4-hydroxyphenylacetate are applied as co-substrates [53]. Some mixture cultures do not need halogenated compounds to stimulate PBDEs transformation [50]. Recently, a lactate-dependent bacterium, *Acetobacterium* sp. strain AG, was isolated and can transform penta-BDE mixtures without other halogenated electron acceptors [52]. We also found that the cometabolism with glucose facilitated the biodegradation of mono-BDE, in terms of kinetics and efficiency in one anaerobic sludge in Taiwan [54].

In a mix microcosm, anaerobic environment necessary for dehalorespiration could be established by other symbiotic microorganisms. In our previous study, the mono-BDE is transformed to diphenyl ether in an aerobic culture from sewage sediment, indicating an anaerobic debromination reaction occurred. The enriched *Clostridiales* specie shown in the denatured gradient gel electrophoresis (DGGE) may responsible for such reaction [55].

2.3. Degradation enzymes

The metabolic pathway of VC is much clearer than that of cDCE. Alkene monooxygenase (AkMO) involves in the initial epoxidation step. The encoded genes (*etnABCD*) and the structures have been identified. Downstream events of the transformation are mediated through coenzyme M transferase (encoded by *etnE* gene), alcohol/aldehyde dehydrogenase, CoA transferase and CoM reductase/carboxylase. The final product, acyl-CoA, is then metabolized through TCA cycle [56]. Proteomic and transcriptomic analyses have confirmed the roles of above enzymes in aerobic VC transformation process.

Aerobic cometabolic degradation of chloroethenes is supposed through several kinds of oxygenases: toluene monooxygenase, toluene dioxygenase, phenol monooxygenase and methane monooxygenase [57]. *P. stutzeri* OX1 depletes PCE and releases chloride irons when toluene is applied as an auxiliary substrate [39]. PCE, DCE, and VC could be transformed by the purified toluene-*o*-xylene monooxygenase (ToMO). ToMO is a four-component enzyme which consists a catalytic oxygen-bridged dinuclear center encoded by *touABE*, a NADH ferredoxin oxidoreductase (encoded by *touF*), a mediating protein (encoded by *touD*), and a Rieske-type ferredoxin (encoded by *touC*). The *touA~F* genes cloned into *E. Coli* could make it to be PCE-degradable.

Different dioxygenases are supposed to involve in aerobic degradation of lower numbered PBDEs. 1,2-dioxygenase is involving in the initial dihydroxylation step when mono-halogenated DEs to be degraded [40]. Downstream degradation processes are supposed through phenol hydroxylases and catechol 1,2-dioxygenase. The transformation by-products range from phenol, catechol, halophenol and halocatechol, indicating nonspecific attack reactions [40, 41]. 2,3-dioxygenase is responsible to dihydroxylate lower numbered PBDEs and their similar chemicals such as DE in the close species *Sphingomonas sp.* PH-07 [42]. The range of PBDEs transformed by *R. jostii* RHA1 depends on the types of growth substrate. The enzymes responsible for degradation are inducible [43]. The expression of biphenyl dioxygenase (BPDO) and ethylbenzene dioxygenase (EBDO) are upregulated during PBDEs degradation. Ectopically expression of these enzymes in closed bacteria that bears no PBDEs degradation activity could transform PBDEs. EBDO depleted mono- through penta-BDEs and BPDO only depleted mono-, di- and one tetra-BDEs. [58].

The structures of HBCD and hexachlorocyclohexanes (HCHs) are quiet similar. Heeb et al. purified the HCH-converting haloalkane dehalogenase LinB, from *Sphigobium indicum* B90A and applied the enzyme for HBCD degradation. LinB transforms HBCD into pentabromocyclododecanols (PBCDOHs) and further tetrabromocyclododecadiols (TBCDDOHs) [59]. Whether LinB or other haloalkane dehalogenase are the de novo HBCD degradation enzyme is unknown. What enzyme responsible for HBCD degradation in *Pseudomonas* sp. HB01 is also waited to be uncovered.

Reductive dechlorination reactions are catalyzed by reductive dehalogenases (RDases) The purified PCE RDase, PceA, has proved to transform PCE and TCE to cDCE [60]. The function of TCE RDase, TceA, in transforming TCE to ethene has also been identified [61]. VcrA and BvcA catalyze the transformation of DCE to ethene [62, 63]. In addition to chloroethenes, RDases also could reduce other chlorinated compounds.

RDases which could debrominate PBDEs have not yet been identified. However, some PBDEs degradation bacteria also could transform chloroethene (Table 2), such as *Dehalococcoides* sp., *Desulfitobacterium* sp., and *Sulfurospirillum* sp. Whether these microorganisms use chloroethene RDases to transform PBDEs is unknown. It is also possible that enzymes with different degradation activities or substrate specificity within single degrader may cooperatively transform different PBDEs congeners.

2.4. The structure and function of reductive dehalogenase

Most RDases presented similar features and conserved motifs [28, 29]. In the N-terminus, RDases possess a putative signal sequence containing the twin-arginine translocation (Tat) motif. Such motif is presented in secretary proteins to be transported across the cytoplasmic membrane through the Tat export system. It is proposed that newly synthesized RDase proteins is folded with cofactors (corrinoid and iron-sulfur clusters) in the cytoplasm with the aids of chaperone proteins. The Tat sequence is then proteolytically cleaved during the maturation process. In the C-terminus, two iron-sulfur cluster binding motifs are presented. The Fe-S clusters cooperate with corrinoid, transfering electrons from upstream donors to chloroethenes and thus catalyze the dehalogenation reaction [28].

http://dx.doi.org/10.5772/56306

The localization of chloroethene RDases is supposed in the membrane, where they could accept electrons from proton producing hydrogenase via menaquinone. The membrane-bound characteristics of RDases has been proved, such as PceA of *D. ethenogenes*. The localization of constitutively expressed PceA in *S. multivorans* was initially found in the cytoplasmic fraction [65]. John et al. used freeze-fracture replica immunogold labeling technique and found it would be at the cytoplasmic membrane when cells grown on pyruvate or formate as electron donors [66].

2.5. Genomic structure and transcription regulation of reductive dehalogenases

The major catalysis reaction of RDases is directed by subunit A, encoded by reductive dehalogenase homologous A (*rdhA*) genes. Over 650 *rdhA* genes have been identified based upon genomic sequence annotation or homologous cloning [67]. However, most of them are not yet been functional characterized. It is common for one dehalorespiration bacterium baring multicopy of *rdhA* genes in the genome. Besides *tceA* and *pceA* genes, there is still 17 RDase genes with unknown function in the genome of *D. ethenogenes* strain 195 [68]. Whether the roles of these genes are relevant to dehalogenation remains unclear.

Most *rdhA* genes are organized with genes encoding for accessory proteins. The *pce*-gene cluster from *D. hafniense* strain Y51 constitutes *pceA* followed by *pceB*, *pceC* and *pceT* [69]. PceT is the trigger factor involving in folding newly synthesized polypeptides. It interacts with the Tat motif of PceA, thus solubilizing and stabilizing PceA polypeptide proceeding downstream maturation and transportation processes [62]. PceB protein contains three transmembrane domains and is assumed as a membrane anchor protein of PceA. PceC contains six-transmembrane domains, an FMN binding domain, and a C-terminal polyferredoxin-like domain. It is similar to the membrane-binding transcription regulators [28, 60]. More examples for the organization of different RDase gene clusters are presented in [28]. *RdhA* and *rdhB* genes usually locate adjacent and are the basic components of the *rdh* gene cluster. They would co-express in order to perform dehalorespiration together.

The expression and silence of RDases during dechlorination reaction is dynamic and regulated. It could be monitored through the amount of RNA or protein. *Dehalococcoides* sp. strain MB bares 7 RDase genes. Only *dceA6* is highly expressed when PCE and TCE are transformed into tDCE. Transcription regulation protein binding site related to gene expression is detected in the upstream of *dceA6* gene [70]. A shotgun metagenome microarray is created to investigate gene transcription in a mixed culture. *rdhA14* and *rdhB14*, are the only two with higher transcript levels during VC degradation, while another 4 *rdhA* genes has higher transcript levels in the absence of VC [71]. The absolute quantification of RDase proteins during the dechlorination process is performed by using nano-liquid chromatography-tandem mass spectrometry in two PCE/TCE degradation consortia. Within 5 selected RDases, only the quantities of PceA and TceA are detectable [72]. The regulation on the expression of *rdh* genes during dechlorination reaction or steady state is not clear. How the physiological environments affect the gene expression is also unclear. Uncover these questions would be helpful for environmental monitoring and remediation.

2.6. The dynamic of degrading population and the evolution of degradation ability

The complete dehalogenation requires different microorganisms which bear various functions in degradation or growth support. Besides, it competes with methanogens and other reducers for H_2. The snapshot of the microbial composition stands for specific ecological condition. The dynamic of composition reveals the effect of various remediation treatments and the interaction between microorganisms. The microbial compositions when co-incubated with zerovalent iron (ZVI) are analyzed by DGGE. The enrichment of iron-reducing bacteria would support the reduction activity of iron for multiple rounds of reactions; the enrichment of nitrate-reducing bacteria also facilitates the cometabolic dehalogenation. These may due to the synergistic effect [54]. Terminal restriction fragment length polymorphism (TRFLP) analysis is also used to analyze the microbial compositions [73]. The resolution limitation of these techniques makes underestimating the complexity of a community. Therefore, new technique is needed to detect specific microbes that are responsible for a key biodegradation process while present in the communities in low numbers. The 16S rRNA genes within a community could be analyzed by recently evolved pyrosequencing or phylogenetic microarray (PhyloChip). PhyloChip composes ten thousands unique 16S rRNA genes. The microbial compositions in TCE contaminated groundwater that is biostimulated or bioaugmented are analyzed. The increase of methanogens at late treatment stage coincident with the increase in methane concentration [74].

There is no close phylogenic relation among diverse dehalorespration degraders. Horizontal gene transfer (HGT) though transposable elements, transmissible plasmids or phage infections is assumed for such convergent evolution. Phylogenetic analyses of the sequences of *rdh* operon and the adjacent genomic structures support HGT. The *pceABC* operon in *D. hafniense* strain TCE1 has been shown to be presented in a circularized transposable element, Tn-Dha1 [75].The single-copy transfer messenger RNA gene (*ssrA*) essential in bacteria is a common target for mobile element. Integration of mobile element results in the duplication of *ssrA* gene around transported gene cluster. Many strain-specific *rdhA* genes collocates within such structures and in a region of high genomic variability between *Dehalococcoides* strains [76]. According to the metagenomic sequence analysis, one prophage element is located adjacent to *tceAB* genes in the *Dehalococcoides*-containing consortium, KB-1. The failure in detecting *tceA* gene expression in virus and the more closed transposase genes indicating higher possibility for HGT through transposable elements [71]. It seems that dehalorespiration degraders do no acquire RDase genes through single way. This would increase the diversity of degraders and function of RDase, which is advantageous for remediation.

3. Integration of biodegradation with other remediation methods

The degradation rate of natural attenuation is slower than chemical or physical treatments. Biostimulation or bioaugement are common strategies for bioremediation of chloroethene [30] Chemical supplements such as potassium permanganate or oxygen injection which can increase oxygen concentration are benefit for microbial dechlorination [77]. Kuo and his collages set biosparging wells for injection substrate and air into TCE contaminated area. Above 95% TCE was removed through cometabolic reactions because the elevated chemical oxygen demand (COD), microbial population, oxidation-reduction potential (ORP) and specific degrading genes after the supplement of substrate [78]. Shortage of auxiliary substrates or accumulation of toxic intermediates also decrease the dehalogenation effect, combined remediation methods may recover the above drawbacks.

Sequential anaerobic/aerobic biodegradation is one of the approaches to accelerate the degradation of recalcitrant halogenated compounds. Anaerobic degraders could target the higher-numbered halogenated compounds. Aerobic degraders only process lower-numbered halogenated compounds. They could transform the by-products produced from anaerobic degradation to antoxic compounds through metabolic or cometabolic reactions. Integration of these two systems makes it possible for complete mineralization.

Chloroethene and PBDEs can be depleted by microscale or nanoscale zerovalent metals [79, 80]. Preliminary dehalogenation of highly halogenated compounds by the reduced metal to generate less halogenated byproducts those are susceptible for microbial degradation. Therefore, the integration of zerovalent metals with biodegradation promotes the dehalogenation efficiency of each type of remediation methods. Reductive debromination of DBDE with nanoscale ZVI (nZVI) results various intermediates ranging from nona-BDEs to tri-BDEs. The known aerobic PBDEs degrader, *Sphingomonas* sp. strain PH-07, which is able to grow in the presence of nZVI, aerobically mineralizes the low brominated-DEs (tri-BDEs – mono-BDEs) from nZVI treatment [81]. The interactions between metals and microbes are complicated and delicate. H_2 generated from the oxidation of metals promotes the growth of some dehalorespiration microbes. Some microorganisms could reduce oxidized metals for multiple runs of reductive reactions or degrade target compounds through cometabolism. Co-incubation with ZVI, microbes in the DBDE-degrading anaerobic sludges hinders the accessibility of MZVI to DBDE and reduced the removal ability in initial stage. However, the synergistic effect in DBDE degradation appears later on. According to the analysis of the microbial community change, co-incubation with MZVI leads to the enrichment of heterotrophic microbial populations bearing nitrate- or iron-reducing activities. The interaction between MZVI and microbes contributed to the synergistic effect [54]. Not only is the growth of microbes affected by metals, but also the expression of functional RDases. Bare nZVI down-regulate the expression of *tceA* and *vcrA* genes while coated particles up-regulate their expression [82]. In addition to the reduced metals, combining the electro-fenton process in aerobic degradation is also a newly evolved and potential way in bioremediation. Application of electrolysis also stimulate the microbial reductive dechlorination and oxidative activities [83].

There are advantages by using each type of remediation approach, while there are also limitations and drawbacks. Combining biodegradation with other abiotic/biotic degradation approaches could overcome their weakness and accelerate the degradation efficiency. The recalcitrant halogenated compounds could be completely mineralized. The impact on environment might also be minimized. Integration of different approaches is a new direction for future investigation.

4. Conclusion

The current knowledge of microbial degradation of chloroethenes, PBDEs and HBCD, has been summarized and reviewed. The biodegradation of these halogenated compounds through aerobic oxidation, aerobic cometabolization, or reductive dehalogenation are introduced. The correspondent enzymes are discussed from the biochemical and molecular point of view. The structure and function of RDases, as well as gene expression regulation and genomic evolution are the major focus. Integration and sequential anaerobic/aerobic biodegradation or (electro)chemical/microbial degradation are suggested for overcoming the disadvantages of single type of treatment. It is possible to completely mineralize these halogenated pollutants by the combination of bio- and abiotic processes and shows promise for site remediation in natural settings and in engineered systems.

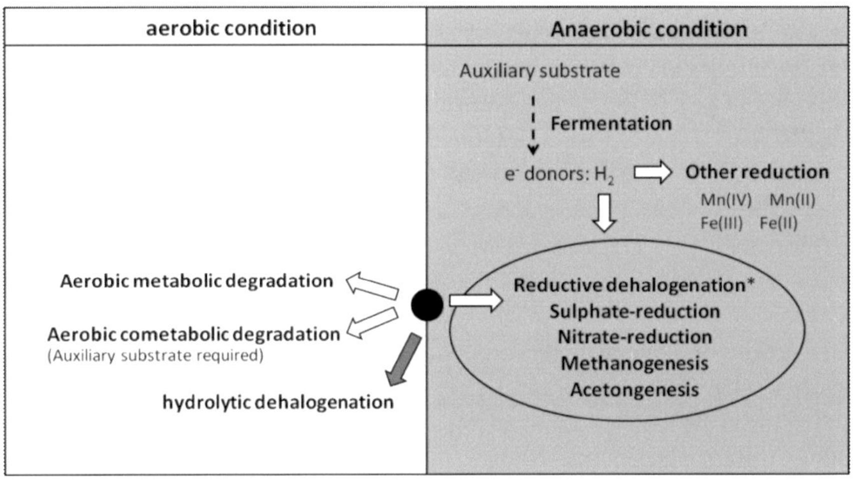

Figure 1. Aerobic degradation and anaerobic reductive dehalogenation reactions of chloeoethenes, PBDEs and HBCD. Circle dot indicates above compounds; star indicates reductive dehalogenation driven by the oxidation of electron donors or occurring cometabolically with other dehalorespiration process; gray arrow indicates the hydrolytic dehalogenation of HBCD.

PCE/TCE	PBDEs	HBCD
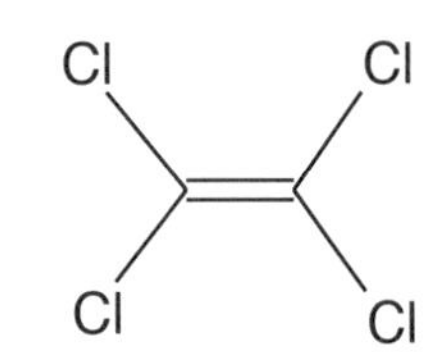 or H	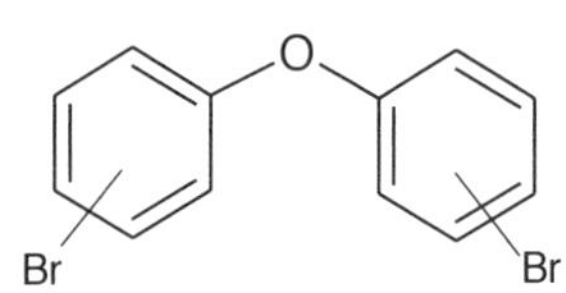	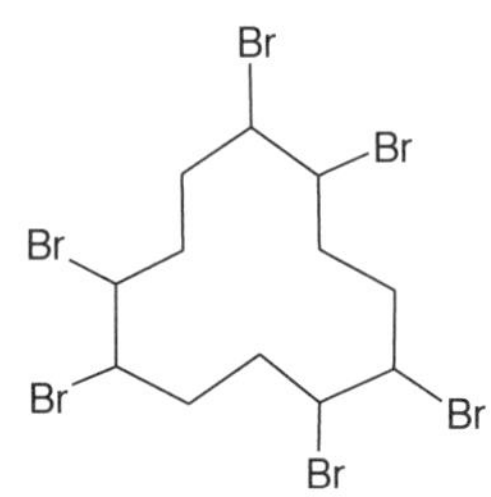

Molecular weight	165/131	249.1 ~ 959.2	641.7
Water solubility (mg l^{-1})	150/1280	4.8 ~0.02	0.003
Toxicity[a]	Epidermis, liver and kidney damage, immune- and neuro-toxicity, reproductive and endocrine effects, probably cancer	Disrupt the balance of thyroid hormone, reproductive, hepatic, and immunotoxicity, developmental neurotoxicity	Thyroidal and developmental toxicity
Abiotic degradation	Chemical oxidation , irradiation, reduced metals (Fe, Fe/Pd)	Pyrolysis, photolysis, reduced metals (Fe, Fe/Pd),	Pyrolysis, photolysis
Biological degradation	Aerobic/anaerobic	Aerobic/anaerobic	Aerobic/anaerobic(?)

a: the toxicity of HBCD is based upon investigation in animal model.

Table 1. Physicochemical properties, biological impacts and degradation routes of chloroethenes, PBDEs and HBCD.

	Substrate	End-products	Genes[c]	references
Aerobic				
Burkholderia xenovorans LB400	Hexa-BDE to mono-BDE,	Hydroxylated-BDE		[43]
Lysinibacillus fusiformis strain DB-1	DBDE	ND		[44]
Pseudonocardia dioxanivorans CB1190	Mono-BDE	ND		[43]
Pseudomonas sp. HB01	HBCD	PBCDOHs, TBCDDOHs		[45]
Pseudomonas stutzeri OX1	PCE	Cl^-,	ToMO-*touABCDEF*	[39]
Rhodococcus jostii RHA1	Penta-BDE to mono-BDE	Hydroxylated-BDE	BPDO- *bphAa*, EBDO- *etbAa1*, *etbAc*	[43, 84]
Rhodococcus sp. RR1	Di-, and mono-BDE	ND		[43]
Sphingomonas sp. PH-07	Tri-, di-, and mono-BDE	Catechol, dibromophenol, dihydroxy mono- and dibromo-BDE		[42]
Sphingomonas sp. SS3	Fluoro-, chloro-, and bromo-DE	Phenol, catechol, Halophenol and Halocatehol		[40]

Sphingomonas sp. SS33	Di-, and mono-; fluoro-, chloro-, and bromo-DE	Phenol, catechol, di-, and mono- halophenol, di-, and mono-halocatehol		[41]
Anaerobic				
Acetobacterium sp. strain AG	Penta-BDE mixture[a]	Tetra-, tri-, and di- BDEs		[52]
Bacillus sp JSK1	PCE	*Cis*-DCE		[85]
Dehalobacter restrictus	PCE, TCE	*Cis*-DCE	*pceA,*	[67, 86]
Dehalococcoides sp. strain BAV1	DCE, VC	Ethene	*bvcA*	[49, 87]
Dehalococcoides sp. strain DG	TCE Octa-BDE mixture[b] Penta- BDE mixture[a]	Ethene Tetra- and penta- BDEs Tetra- BDE		[52]
Dehalococcoides ethenogenes strain 195	PCE, TCE, cis -DCE, and VC Octa-BDE mixture[b]	Ethene Tetra-, penta-, hexa- and Penta-BDEs	*pceA, tceA*	[51, 67, 89]
Dehalococcoides sp. strain MB	PCE, TCE	*Trans*-DCE		[90]
Desulfitobacterium chlororespirans strain Co23	Octa-BDE	ND		[53]
Desulfitobacterium dehalogenans strain JW/IU-DC1	PCE, TCE, Octa-BDE,	ND		[53, 67, 91]
Desulfitobacterium hafniense PCP-1	Octa-BDE	ND		[53]
Geobacter lovleyi strain SZ	PCE, TCE	cDCE		[92]
Sulfurospirillum multivorans	PCE, TCE DBDE	DCE Octa- and hepta- BDEs	*pceA*	[51, 67]

a penta- BDE mixture: hexa-, penta-, and tera-BDEs.

b Octa-BDE mixture: nona-, octa-, hepta-, and hexa-BDEs.

c Genes that only relevant to degradation of chloroethens, PBDEs and HBCD are listed.

ND: data not shown. BPDO: biphenyl dioxygenase. EBDO: ethylbenzene dioxygenase. PBCDOHs: pentabromocyclododecanols. TBCDDOHs: tetrabromocyclododecadiols.

Table 2. Selected bacteria which degrade chloroethens, PBDEs and HBCD.

http://dx.doi.org/10.5772/56306

Acknowledgements

The authors thank National Science Council (NSC), Taiwan, ROC for financial support.

Author details

Yu-Huei Peng and Yang-hsin Shih*

*Address all correspondence to: yhs@ntu.edu.tw

Department of Agricultural Chemistry, National Taiwan University, Taipei, Taiwan

References

[1] Toxicological Profile for Tetrachloroethylene (Update). ATSDR 1997.

[2] Tetrachloroethylene. IARC monograph 2007.

[3] Toxicological Profile for Trichloroethylene (Update). ATSDR 1997.

[4] Toxicologica Review of Vinyl Chloride. USEPA 2000.

[5] 1,2-dichloroethene. ATSDR 1997.

[6] Costa L. G., Giordano G. Developmental neurotoxicity of polybrominated diphenyl ether (PBDE) flame retardants. Neurotoxicology 2007; 28(6) 1047-1067.

[7] Tseng L. H. et al. Developmental exposure to decabromodiphenyl ether (PBDE 209): effects on thyroid hormone and hepatic enzyme activity in male mouse offspring. Chemosphere 2008; 70(4) 640-647.

[8] An J. et al. The cytotoxic effects of synthetic 6-hydroxylated and 6-methoxylated polybrominated diphenyl ether 47 (BDE47). Environmental Toxicology 2011; 26(6) 591-599.

[9] Yan C. et al. The involvement of ROS overproduction and mitochondrial dysfunction in PBDE-47-induced apoptosis on Jurkat cells. Experimental and Toxicologic Pathology 2011; 63(5) 413-417.

[10] Gauthier L. T. et al. Temporal trends and spatial distribution of non-polybrominated diphenyl ether flame retardants in the eggs of colonial populations of Great Lakes herring gulls. Environmental Science & Technology 2009; 43(2) 312-317.

[11] Marvin C. H. et al. Hexabromocyclododecane: Current Understanding of Chemistry, Environmental Fate and Toxicology and Implications for Global Management. Environmental Science & Technology 2011; 45(20) 8613-8623.

[12] Deng J. et al. Hexabromocyclododecane-induced developmental toxicity and apoptosis in zebrafish embryos. Aquat Toxicol 2009; 93(1) 29-36.

[13] HBCD Factsheet. BSEF 2009.

[14] Hexabromocyclododecane (HBCD) Action Plan. USEPA (2010).

[15] Summary of a Proposal to List Hexabromocyclododecane in Annex A to the Convention. POPRC 2009.

[16] Honning J. et al. Role of diffusion in chemical oxidation of PCE in a dual permeability system. Environmental Science & Technology 2007; 41(24) 8426-8432.

[17] Weber R. et al. PCB destruction in subcritical and supercritical water--evaluation of PCDF formation and initial steps of degradation mechanisms. Environmental Science & Technology 2002; 36(8) 1839-1844.

[18] Lin Y. M. et al. Emissions of Polybrominated Diphenyl Ethers during the Thermal Treatment for Electric Arc Furnace Fly Ash. Aerosol and Air Quality Research 2012; 12(2) 237-250.

[19] Barontini F. et al. Thermal stability and decomposition products of hexabromocyclododecane. Industrial & Engineering Chemistry Research 2001; 40(15) 3270-3280.

[20] Shih Y.-h., Wang C.-K. Photolytic degradation of polybromodiphenyl ethers under UV-lamp and solar irradiations. Journal of Hazardous Materials 2009; 165(1-3) 34-38.

[21] Beitinger E. Permeable treatment walls - Design, construction, and cost. in: NATO/CCMS Pilot Study: Evaluation of Demonstrated and Emerging Technologies for the Treatment of Contaminated land and Grooundwater (Phase III). 1998.

[22] He F., Zhao D. Y. Hydrodechlorination of trichloroethene using stabilized Fe-Pd nanoparticles: Reaction mechanism and effects of stabilizers, catalysts and reaction conditions. Applied Catalysis B-Environmental 2008; 84(3-4) 533-540.

[23] Shih Y. H., Tai Y. T. Reaction of decabrominated diphenyl ether by zerovalent iron nanoparticles. Chemosphere 2010; 78(10) 1200-1206.

[24] Su Y.-f. et al. Effects of various ions on the dechlorination kinetics of hexachlorobenzene by nanoscale zero-valent iron. Chemosphere 2012; 88(11) 1346-1352.

[25] Shih Y.-h. et al. Pentachlorophenol reduction by Pd/Fe bimetallic nanoparticles: Effects of copper, nickel, and ferric cations. Applied Catalysis B: Environmental 2011; 105(1-2) 24-29.

[26] Shih Y. H. et al. Reduction of hexachlorobenzene by nanoscale zero-valent iron: kinetics, pH effect, and degradation mechanism. Separation and purification technology (2011)., 76(3), 268-274.

[27] Lee C. et al. Bactericidal effect of zero-valent iron nanoparticles on Escherichia coli. Environmental Science & Technology 2008; 42(13) 4927-4933.

[28] Futagami T. et al. Biochemical and genetic bases of dehalorespiration. Chemical record 2008; 8(1) 1-12.

[29] Smidt H., de Vos W. M. Anaerobic microbial dehalogenation. Annual review of microbiology (2004)., 58, 43-73.

[30] Tiehm A., Schmidt K. R. Sequential anaerobic/aerobic biodegradation of chloroethenes - aspects of field application. Current Opinion in Biotechnology 2011; 22(3) 415-421.

[31] Huang H. et al. In vitro biotransformation of PBDEs by root crude enzyme extracts: Potential role of nitrate reductase (NaR) and glutathione S-transferase (GST) in their debromination. Chemosphere (2012)., 90(6), 1885-1892.

[32] Davis J. W. et al. The transformation of hexabromocyclododecane in aerobic and anaerobic soils and aquatic sediments. Water Research 2005; 39(6) 1075-1084.

[33] Hakk H., Letcher R. J. Metabolism in the toxicokinetics and fate of brominated flame retardants--a review. Environmental international 2003; 29(6) 801-828.

[34] Tiehm A. et al. Growth kinetics and stable carbon isotope fractionation during aerobic degradation of cis-1,2-dichloroethene and vinyl chloride. Water Research 2008; 42(10–11) 2431-2438.

[35] Olaniran A. O. et al. Aerobic biodegradation of dichloroethenes by indigenous bacteria isolated from contaminated sites in Africa. Chemosphere 2008; 73(1) 24-29.

[36] Schmidt K. R. et al. Aerobic biodegradation of cis-1,2-dichloroethene as sole carbon source: Stable carbon isotope fractionation and growth characteristics. Chemosphere 2010; 78(5) 527-532.

[37] Zhang Y., Tay J. H. Co-metabolic degradation activities of trichloroethylene by phenol-grown aerobic granules. Journal of Biotechnology 2012; 162(2-3) 274-282.

[38] Frascari D. et al. A kinetic study of chlorinated solvent cometabolic biodegradation by propane-grown Rhodococcus sp. PB1. Biochemical Engineering Journal 2008; 42(2) 139-147.

[39] Ryoo D. et al. Aerobic degradation of tetrachloroethylene by toluene-o-xylene monooxygenase of Pseudomonas stutzeri OX1. Nat Biotech 2000; 18(7) 775-778.

[40] Schmidt S. et al. Biodegradation of diphenyl ether and its monohalogenated derivatives by Sphingomonas sp. strain SS3. Applied and Environmental Microbiology 1992; 58(9) 2744-50.

[41] Schmidt S. et al. Biodegradation and Transformation of 4,4'-Dihalodiphenyl and 2,4-Dihalodiphenyl Ethers by Sphingomonas Sp Strain Ss33. Applied and Environmental Microbiology 1993; 59(11) 3931-3933.

[42] Kim Y. M. et al. Biodegradation of diphenyl ether and transformation of selected brominated congeners by Sphingomonas sp. PH-07. Applied microbiology and biotechnology 2007; 77(1) 187-194.

[43] Robrock K. R. et al. Aerobic biotransformation of polybrominated diphenyl ethers (PBDEs) by bacterial isolates. Environmental Science & Technology 2009; 43(15) 5705-5711.

[44] Deng D. et al. Aerobic debromination of deca-BDE: Isolation and characterization of an indigenous isolate from a PBDE contaminated sediment. International Biodeterioration & Biodegradation 2011; 65(3) 465-469.

[45] Yamada T. et al. Isolation of Pseudomonas sp. strain HB01 which degrades the persistent brominated flame retardant gamma-hexabromocyclododecane. Bioscience, Biotechnology, and Biochemistry 2009; 73(7) 1674-1678.

[46] Aulenta F. et al. Competition for H2 between sulfate reduction and dechlorination in butyrate-fed anaerobic cultures. Process Biochemistry (2008)., 43(2), 161-168.

[47] Conrad M. E. et al. Field evidence for co-metabolism of trichloroethene stimulated by addition of electron donor to groundwater. Environmental Science & Technology (2010)., 44(12), 4697-4704.

[48] John M. et al. Retentive Memory of Bacteria: Long-Term Regulation of Dehalorespiration in Sulfurospirillum multivorans. Journal of Bacteriology 2009; 191(5) 1650-1655.

[49] He J. Z. et al. Detoxification of vinyl chloride to ethene coupled to growth of an anaerobic bacterium. Nature 2003; 424(6944) 62-65.

[50] Lee L. K., He J. Reductive Debromination of Polybrominated Diphenyl Ethers by Anaerobic Bacteria from Soils and Sediments. Applied and environmental microbiology 2010; 76(3) 794-802.

[51] He J. Z. et al. Microbial reductive debromination of polybrominated diphenyl ethers (PBDEs). Environmental Science & Technology 2006; 40(14) 4429-4434.

[52] Ding C. et al. Isolation of Acetobacterium sp. strain AG that reductively debrominates octa- and penta-brominated diphenyl ether technical mixtures. Applied and Environmental Microbiology (2012)., doi:10.1128/AEM.02919-12

[53] Robrock K. R. et al. Pathways for the anaerobic microbial debromination of polybrominated diphenyl ethers. Environmental Science & Technology 2008; 42(8) 2845-52.

[54] Shih Y. H. et al. Synergistic effect of microscale zerovalent iron particles combined with anaerobic sludges on the degradation of decabromodiphenyl ether. Bioresource Technology (2012)., 108, 14-20.

[55] Chen C. Y. et al. Microbial degradation of 4-monobrominated diphenyl ether in an aerobic sludge and the DGGE analysis of diversity. Journal of Environmental Science and Health Part B-Pesticides Food Contaminants and Agricultural Wastes 2010; 45(5) 379-385.

[56] Mattes T. E. et al. Aerobic biodegradation of the chloroethenes: pathways, enzymes, ecology, and evolution. FEMS Microbiology Reviews 2010; 34(4) 445-475.

[57] Furukawa K. Oxygenases and dehalogenases: molecular approaches to efficient degradation of Chlorinated environmental pollutants. Bioscience, Biotechnology, and Biochemistry 2006; 70(10) 2335-2348.

[58] Robrock K. R. et al. Biphenyl and Ethylbenzene Dioxygenases of Rhodococcus jostii RHA1 Transform PBDEs. biotechnology and bioengineering 2011; 108(2) 313-321.

[59] Heeb N. V. et al. Biotransformation of Hexabromocyclododecanes (HBCDs) with LinB—An HCH-Converting Bacterial Enzyme. Environmental Science & Technology 2012; 46(12) 6566-6574.

[60] Suyama A. et al. Molecular Characterization of the PceA Reductive Dehalogenase of Desulfitobacterium sp. Strain Y51. Journal of Bacteriology 2002; 184(13) 3419-3425.

[61] Magnuson J. K. et al. Trichloroethene reductive dehalogenase from Dehalococcoides ethenogenes: Sequence of tceA and substrate range characterization. Applied and Environmental Microbiology 2000; 66(12) 5141-5147.

[62] Maillard J. et al. Redundancy and specificity of multiple trigger factor chaperones in Desulfitobacteria. Microbiology-Sgm (2011)., 157, 2410-2421.

[63] Tang S. et al. Functional Characterization of Reductive Dehalogenases Using Blue Native Polyacrylamide Gel Electrophoresis. Applied and Environmental Microbiology (2012)., 79(3), 974-981.

[64] Maillard J. et al. Characterization of the corrinoid iron-sulfur protein tetrachloroethene reductive dehalogenase of Dehalobacter restrictus. Applied and Environmental Microbiology 2003; 69(8) 4628-4638.

[65] John M. et al. Growth substrate dependent localization of tetrachloroethene reductive dehalogenase in Sulfurospirillum multivorans. Archives of Microbiology 2006; 186(2) 99-106.

[66] Regeard C. et al. Development of degenerate and specific PCR primers for the detection and isolation of known and putative chloroethene reductive dehalogenase genes. J Microbiol Methods 2004; 56(1) 107-18.

[67] West K. A. et al. Comparative genomics of "Dehalococcoides ethenogenes" 195 and an enrichment culture containing unsequenced "Dehalococcoides" strains. Applied and Environmental Microbiology 2008; 74(11) 3533-3540.

[68] Furukawa K. et al. Biochemical and molecular characterization of a tetrachloroethene dechlorinating Desulfitobacterium sp strain Y51: a review. Journal of Industrial Microbiology & Biotechnology 2005; 32(11-12) 534-541.

[69] Chow W. L. et al. Identification and transcriptional analysis of trans-DCE-producing reductive dehalogenases in Dehalococcoides species. ISME journal 2010; 4(8) 1020-1030.

[70] Waller A. S. et al. Transcriptional Analysis of a Dehalococcoides-Containing Microbial Consortium Reveals Prophage Activation. Applied and Environmental Microbiology 2012; 78(4) 1178-1186.

[71] Werner J. J. et al. Absolute quantification of Dehalococcoides proteins: enzyme bioindicators of chlorinated ethene dehalorespiration. Environmental microbiology 2009; 11(10) 2687-2697.

[72] Révész S. et al. Bacterial community changes in TCE biodegradation detected in microcosm experiments. International Biodeterioration & Biodegradation 2006; 58(3–4) 239-247.

[73] Lee P. K. H. et al. Phylogenetic Microarray Analysis of a Microbial Community Performing Reductive Dechlorination at a TCE-Contaminated Site. Environmental Science & Technology 2011; 46(2) 1044-1054.

[74] Maillard J. et al. Isolation and characterization of Tn-Dha1, a transposon containing the tetrachloroethene reductive dehalogenase of Desulfitobacterium hafniense strain TCE1. Environmental microbiology 2005; 7(1) 107-117.

[75] Sahl J. W. et al. Coupling permanganate oxidation with microbial dechlorination of tetrachloroethene. Water Environ Res. 2007; 79(1) 5-12.

[76] Kuo Y. C. et al. Remediation of TCE-contaminated groundwater using integrated biosparging and enhanced bioremediation system. Research Journal of Chemistry and Environment 2012; 16(2) 37-47.

[77] Liu Y. Q. et al. TCE dechlorination rates, pathways, and efficiency of nanoscale iron particles with different properties. Environmental Science & Technology 2005; 39(5) 1338-1345.

[78] Keum Y. S., Li Q. X. Reductive debromination of polybrominated diphenyl ethers by zerovalent iron. Environmental Science & Technology 2005; 39(7) 2280-2286.

[79] Kim Y.-M. et al. Degradation of polybrominated diphenyl ethers by a sequential treatment with nanoscale zero valent iron and aerobic biodegradation. Journal of Chemical Technology & Biotechnology 2012; 87(2) 216-224.

[80] Xiu Z. M. et al. Effect of Bare and Coated Nanoscale Zerovalent Iron on tceA and vcrA Gene Expression in Dehalococcoides spp. Environmental Science & Technology 2010; 44(19) 7647-7651.

[81] Lohner S. T., Tiehm A. Application of Electrolysis to Stimulate Microbial Reductive PCE Dechlorination and Oxidative VC Biodegradation. Environmental Science & Technology 2009; 43(18) 7098-7104.

[82] Robrock K. R., Mohn, W.W., Eltis, L.D., Alvarez-Cohen, L. Biphenyl and ethylbenzene dioxygenases of Rhodococcus jostii RHA1 transform PBDEs. biotechnology and bioengineering 2010; 108(2) 313-321.

[83] Kalimuthu K. et al. Reductive dechlorination of perchloroethylene by bacillus sp JSK1 isolated from dry cleaning industrial sludge. Carpathian Journal of Earth and Environmental Sciences 2011; 6(1) 165-170.

[84] Holliger C. et al. A highly purified enrichment culture couples the reductive dechlorination of tetrachloroethene to growth. Applied and Environmental Microbiology 1993; 59(9) 2991-2997.

[85] Krajmalnik-Brown R. et al. Genetic identification of a putative vinyl chloride reductase in Dehalococcoides sp. strain BAV1. Applied and Environmental Microbiology 2004; 70(10) 6347-51.

[86] MaymoGatell X. et al. Isolation of a bacterium that reductively dechlorinates tetrachloroethene to ethene. Science 1997; 276(5318) 1568-1571.

[87] Cheng D., He J. Z. Isolation and Characterization of "Dehalococcoides" sp Strain MB, Which Dechlorinates Tetrachloroethene to trans-1,2-Dichloroethene. Applied and Environmental Microbiology 2009; 75(18) 5910-5918.

[88] Villemur R. et al. Occurrence of several genes encoding putative reductive dehalogenases in Desulfitobacterium hafniense/frappieri and Dehalococcoides ethenogenes. Canadian journal of microbiology 2002; 48(8) 697-706.

[89] Sung Y. et al. Geobacter lovleyi sp. nov. Strain SZ, a Novel Metal-Reducing and Tetrachloroethene-Dechlorinating Bacterium. Applied and Environmental Microbiology 2006; 72(4) 2775-2782.

Chapter 5

Biodegradation of Melanoidin from Distillery Effluent: Role of Microbes and Their Potential Enzymes

Anita Rani Santal and Nater Pal Singh

Additional information is available at the end of the chapter

http://dx.doi.org/10.5772/56252

1. Introduction

Bioremediation is the process that deals with the microbial degradation of hazardous compounds from environment. The process of bioremediation is the natural process of biodegradation, which can degrade the pollutants and sometimes can completely oxidize the compound. Microorganisms play the vital role in the process of bioremediation and biodegradation because of their great metabolic diversity, which includes the ability to metabolise these pollutants [1]. The degradation of toxic compounds to less harmful forms with the use of biological systems is called as bioremediation [2]. Bioremediation is limited in the number of toxic material, it can handle but where applicable it is cost effective and ecofriendly [3]. Today, water resources have been the most exploited of the natural systems, most of our water bodies are seriously polluted due to rapid population growth, industrial proliferation, urbanizations, increasing living standards and wide spheres of human activities. Many rivers of the world receive heavy flux due to industrial effluents [4]. The wastewater consisting of substances varying from simple nutrients to highly toxic hazardous chemicals, which when used for irrigation caused both beneficial and damaging effects to various crops including vegetables [5].

The awful consequence of increasing population along with an rapidly increase in the industrial field is a major concern of pollution in the environment. Water is necessary for irrigation to increase the productivity in agriculture. Industrial effluents already pollute most of the Indian rivers. Many industries are playing the crucial role in water pollution such as textile industries, dairy industries and distillery *etc*. Distilleries are the major agro-based industries, which utilize molasses as raw material for the production of rectified spirit. In addition to rectified spirit, distilleries also produce ethanol, which can be mixed with diesel and used as biofuel and helps in reducing import of crude oil thereby saving foreign exchange

[6]. In the year 1999, there were 285 distilleries in India producing 2.70×10^9 L of wastewater each year [7]. This number has gone up to 319 producing 3.25×10^9 L of alcohol and generating 4.04×10^{11} L of wastewater annually [8] and this effluent or spent wash is a major source of aquatic and soil pollution [9]. The spentwash is highly acidic in nature and has a variety of recalcitrant coloring compounds as melanoidins, phenolics and metal sulfides that are mainly responsible for the dark color of distillery effluent. The pH of spentwash increases from 4.5 to 8.5 during the anaerobic treatment process and finally it is called as post methanated distillery effluent (PMDE). The spentwash is a waste having very high biochemical oxygen demand (BOD; 35,000-40,000 mgL^{-1}), chemical oxygen demand (COD; 90,000-1,10,000 mgL^{-1}), total solids (TS; 82,480 mgL^{-1}), nitrogen (2,200 mgL^{-1}), phenolics (4.20 mgL^{-1}) and sulphate (3,410 mgL^{-1}). In addition to these contaminants, several heavy metals (Cd, Mn, Fe, Zn, Ni and Pb) are also present. The spentwash primarily undergoes anaerobic treatment process which converts a significant proportion (>50%) of the BOD and COD. However, different biochemical changes in the spentwash occur during anaerobic digestion. There is formation of significant amount of hydrogen sulfide (H_2S) as a result of the reduction of oxidized sulfur compound. Sulfide binds with the heavy metals present in the effluent and forms a colloidal solution of metal sulfide colorant [10]. This massive quantity, approx. 40 billion litres of effluent, if disposed untreated can cause considerable stress on the watercourses leading to widespread damage to aquatic life. Moreover, the oxygen is also exhausted giving death of fishes and other aquatic life. This wastewater is extremely harmful to the plants. Cane molasses contained around 2% of dark brown pigment melanoidins [11]. Dark brown color hinders photosynthesis by blocking sunlight and is therefore deleterious to aquatic life [12-14]. It reduces soil alkalinity and inhibits seed germination. Further due to the possibility of complexation reaction of introduced melanoidins with metal ions, they could influence the biogeochemical cycle of many constituents in natural water [14], which are highly resistant to microbial attack. Hence, the wastewater requires pretreatment before its safe disposal into the environment [15, 16].

This book chapter contributes and summarizes the literature available on the structure, nature and properties of melanoidins. Besides these, the important part of this chapter involves role of bacteria, fungi, yeast and algae helps in biodegradation of melanoidins and also the enzymes involved in the degradation of melanoidins. Therefore, this chapter has a major significance in the prevention of environmental pollution.

2. Nature of melanoidins

Melanoidins are natural, dark brown, complex biopolymers produced by non-enzymatic browning reactions called as Maillard amino-carbonyl reaction taking place between the amino and carbonyl groups in organic substances [17, 18]. It is considered that melanoidins extensively exist in food, drinks and wastewater released from distillery and fermentation industries. Melanoidins exist not only in various foods but also in various industrial wastes e.g. distillery and sugar mill wastes [16]. Melanoidins owing to their structural complexity, dark color and offensive odour pose threat to aquatic and terrestrial ecosystem. Currently, the most visible environmental problem caused by contamination with melanoidin compounds is eutrophication in natural water bodies, reduction of sunlight penetration leading to decreased photosynthetic activity and dissolved oxygen concentration in lakes, rivers or lagoons [19, 20]. Therefore, the degradation and decolorization of molasses melanoidin by chemical [21] and biological means [20, 22] has been attempted in order to characterize the chemical structure and bioremediation aspects.

Melanoidins are generated through the maillard reaction. Maillard reaction is known to proceed non-enzymatically between amino and carbonyl groups of organic matter [13, 14, 17, 18, 23] and are closely related to humic substances in the natural environment [24]. There are two important pathways in the non-enzymatic browning of food i.e. caramelization and maillard reaction. That leads to the formation of volatile aromatic compounds, intermediate nonvolatile compounds, and brown pigments called melanoidins. Since they are highly resistant to microbial attack, the conventional biological processes are inapplicable to color removal from melanoidin containing wastewater from distillery [13].

Melanoidins have physiologically positive effects such as anti-oxidative activity including strong scavenging activity against reactive oxygen species [13, 25]. Miyata *et al.* [26] in their studies attempt to utilize the strong oxidizing activity of a white rot fungus *Coriolus hirsutus* to decolorise melanoidins, which is a main color component of heat treatment liquor of sewage sludge. This reaction occurs in many situations, in foods [27], *in vivo*, and in soil. Melanoidins, which are the final by-products of the reaction, are brown nitrogen containing polymers that are difficult to decompose. The formation of melanoidins is affected by the reactants and their concentrations, types of catalysts and buffers, reaction temperature, time, pH value, water activity, presence of oxygen and metal ions. During heat treatment, the maillard reaction accompanied by formation of a class of compounds known as maillard products. The reaction proceeds effectively >50 ºC and is favoured at pH 4 to 7 [28]. Recently, the empirical formula of melanoidins has been suggested as $C_{17\text{-}18}H_{26\text{-}27}O_{10}N$ [13]. The basic structure of melanoidin is given in Figure 1 [29]. The molecular weight distribution is between 5-40 kDa. It consists of acidic, polymeric and highly dispersed colloids, which are negatively charged due to the dissociation of carboxylic acids and phenoilc groups [30]. Due to their antioxidant properties, melanoidins are toxic to many microorganisms involved in wastewater treatment [31]. Melanoidins and their degradation are generally recalcitrant to biodegradation due to their complex structure and xenobiotic nature and in some cases are both mutagenic and carcinogenic [14]. Melanoidins behave as anionic hydrophilic polymers, which can form stable complexes with metal cations and reported that ketone or hydroxyl groups of pyranone or pyridone residues act as donor groups in melanoidins and participate in the chelation with metals as melanoidins have net negative charge and therefore, different heavy metals (Cu^{2+}, Cr^{3+}, Fe^{3+}, Zn^{2+}, Pb^{2+}, *etc.*) form large complex molecules with melanoidins, amino acids, proteins and sugars in acidic medium and get precipitated [14, 32].

Miura *et al.* [33] have reported the formation of colored compounds formed in the maillard reaction between D-xylose and glycine. These pigments turned brown with decomposition, so they had been postulated to be important intermediates in the generation of melanoidins. Hayase *et al.* [34] have reported the identification of the novel blue pigment designated as Blue-M1. These consist of four molecules of D-xylose and glycine, and also have methine proton

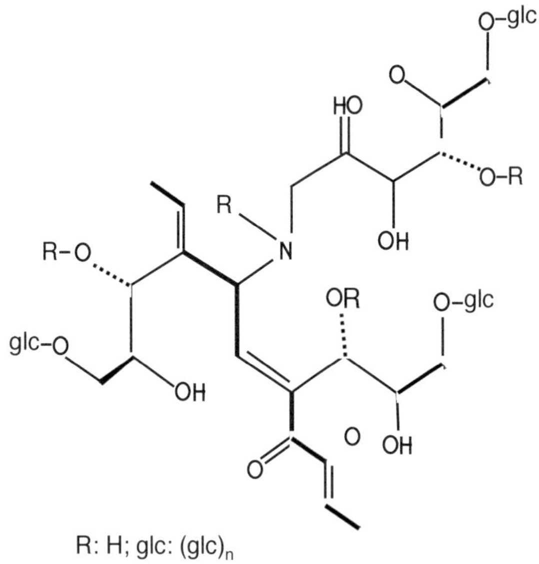

Figure 1. Basic structure of melanoidins [29].

between two pyrrolpyrrole rings. In the initial stage of maillard reaction, disaccharides form a Schiff base with amino groups of amino acids and are subsequently transformed *via* the Amadori rearrangement product [35, 36].

3. Degradation of melanoidin polymer

Waste treatment methods aim the removal of unwanted compounds in wastewater for safe discharge into environment. Various technologies have been used for reducing the pollution load from distillery wastewater, which included physico-chemical treatment (Adsorption, Coagulation and flocculation, Oxidation process, Membrane treatment, Evaporation/combustion) and biological treatment (Figure 2) [37].

Adsorption techniques for wastewater treatment have become more popular in recent years. Among the physico-chemical processes, adsorption technology is considered to be one the most effective and proven technology having potential application in both water and wastewater treatment. Adsorption is a rapid phenomenon of passive sequestration and separation of adsorbate from aqueous/gaseous phase on to solid phase [38]. Decolorization of synthetic melanoidin using commercially available activated carbon as well as activated carbon produced from sugarcane bagasse was investigated by Bernardo *et al.* [39]. Activated carbon is a well known adsorbent due to its extended surface area, microporous structure, high adsorption capacity and high degree of surface reactivity. Previous studies on decolorization of molasses wastewater include adsorption on commercial as well as indigenously prepared activated carbons [40]. Coagulation involves the destabilization of colloidal particles (e.g.

http://dx.doi.org/10.5772/56252

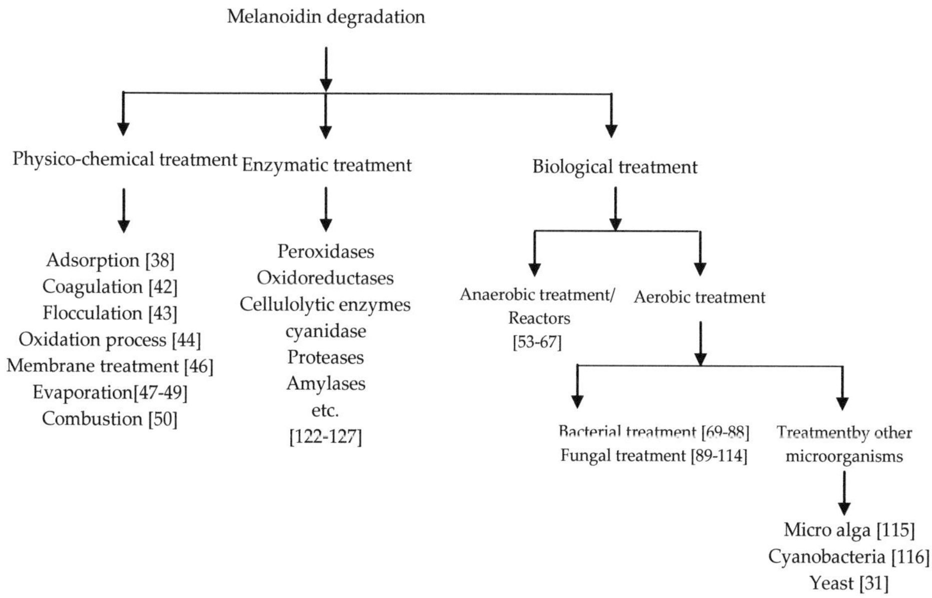

Figure 2. Various methods of melanoidin degradation.

mineral colloids, microbial cells, virus particles) and sometimes by coagulant aids (e.g. activated silica, bentonite, polyelectrolytes, starch) [41]. Coagulation of the wastewater containing recalcitrant pollutants is practiced through use of conventional salts of aluminium and iron as coagulants [42]. Migo *et al.* [43] used a commercial inorganic flocculent, a polymer of ferric hydroxysulfate with a chemical formula $[Fe_2 (OH)_n (SO4)_{3-n/2}]_m$ for the treatment of molasses wastewater.

A study comparing the efficiency in decolorising biologically pretreated molasses wastewater of different oxidation processes using ozone, single hydrogen peroxide, Fenton's reagent and ozone combined with hydrogen peroxide has been performed. Ozone treatment was able to reduce about 76% of color. A combination of ozone with a low concentration of hydrogen peroxide was able to increase the color removal efficiency up to 89%. Gel permeation chromatography corroborated the reduction in the concentration of chromophore groups responsible for wastewater color. Single hydrogen peroxide and Fenton's reagent were not able to reduce the color. Bicarbonate ions were found to be strong inhibitors of decolorising reactions [44]. Kumaresan *et al.* [45] employed emulsion liquid membrane (ELM) technique in a batch process for spentwash treatment. Water-oil-water type of emulsion was used to separate and concentrate the solutes resulting in 86% and 97% decrease in COD and BOD, respectively. Electrodialysis has been explored for desalting spentwash using cation and anion exchange membranes resulting in 50-60% reduction in potassium content [46]. During treatment of winery and distillery wastewater by natural evaporation in ponds, formation of malodorous compounds induces harmful olfactory effects. The organic compounds are oxidised to CO_2 and the nitrate is reduced to N_2 (odourless compounds), without VFA formation. The preventive treatment of odours by nitrate addition was tested on an industrial scale in winery and distillery ponds [47]. Molasses spentwash containing 4% solids can be concentrated to a maximum of 40% solids in a quintuple-effect evaporation system with thermal vapor recompression [48, 49]. Combustion is also an effective method of on-site vinasse disposal as it is accompanied by production of potassium-rich ash [50] that can be used for land application.

Although many techniques have been explored for the treatment of melanoidins but they all are cost effective and produce large amount of sludge. So, to overcome these problem biological methods plays a crucial role in treatment of the melanoidins.

4. Biological methods for the treatment of distillery effluent

Microorganisms (bacteria/fungi/actinomycetes) due to their inherent capacity to metabolize a variety of substrate have been utilized since long back for biodegradation of complex, toxic and recalcitrant compounds, which cause severe damage to environment. Thus, these organisms have been exploited for biodegradation and decolorization of melanoidin pigment present in industrial wastes especially from distillery and fermentation industry. Biological treatment of molasses wastewater undergoes two subdivisions which include anaerobic treatment and aerobic treatment.

Anaerobic treatment is an accepted practice, and various high rate reactor designs have been tried at pilot and full scale operation. Aerobic treatment of anaerobically treated effluent using different microbial populations has also been explored. Majority of biological treatment technologies remove color by either concentrating the color into sludge or by partial or complete breakdown of the color molecules. These methods are discussed in detail in the following section.

4.1. Anaerobic treatment /reactors

In the anaerobic treatment of wastewater, the energy is extracted from the waste components without the introduction of air or oxygen. The anaerobic treatment of wastewater has now emerged as an energy saving wastewater treatment technology. It involves biological processes that take place in the absence of oxygen and in which organic materials are degraded to produce biogas [51]. Molasses wastewater treatment using anaerobic process is a very promising re-emerging technology, which presents interesting advantages as compared to classical aerobic treatment. It produces very little sludge, requires less energy and can be successfully operated at high organic loading rates; also, the biogas thus generated can be utilized for steam generation in the boilers thereby meeting the energy demands of the unit [52].

In India, the traditional method for this type of wastewater treatment involved sequestration in anaerobic lagoons during the dry season, followed by release in to the river. Anaerobic lagoons afford the simplest form of treatment to distillery effluent. Drawbacks of such form of treatment are the requirement of large land area, odour nuisance, and chances of ground water pollution. Since the development of the UASB process in the 1970s, this process has been

http://dx.doi.org/10.5772/56252

widely applied for the treatment of industrial effluents. Effluents from alcohol producing industries are mostly highly polluted and therefore in principle very suitable for anaerobic treatment. Distilleries use different kinds of raw materials such as sugar cane juice, sugar cane molasses, sugar beet molasses, wine or corn for the production of alcohol [40]. The use of different materials and the different processes applied, resulted the production of wide variety of effluent. The process conditions under which good results of the anaerobic process are obtained depend heavily on the type of distillery effluent being treated. The choice of the right set of process parameters for every type of distillery effluent has shown to be of crucial importance for the anaerobic process. Bhoi *et al.* [53] worked on treatability studies of black liquor by upflow anaerobic sludge blanket reactor. This was deals with the feasibility studies of anaerobically treated black liquor along with the readily degradable synthetic wastewater (SWW) in a bench-scale upflow anaerobic sludge blanket reactor (UASBR). Treatment has been carried out in four phases of operation. In the first phase, 100% of SWW prepared from molasses was used. In the other phases of operation, 10%, 20%, and 30% of the SWW chemical oxygen demand (COD) was replaced by Black liquor COD successively. In each of the operation phases, the composition of the feed was not changed until COD reduction and gas production were stabilized. The entire operation took 509 days. The results indicate that the COD removal efficiency exhibited a decreasing trend with increase in the percentage of black liquor and apparently, UASB cannot be used as a treatment option for SWW containing more than 20% black liquor. Farhadian *et al.* [54] studied the treatment of strong beet sugar wastewater by an upflow anaerobic fixed bed (UAFB) at pilot plant scale. Reactors filled with standard industrial packing and inoculated with anaerobic culture (chicken manure, cow manure, anaerobic sludge digested from domestic wastewater) at 32-34 °C with 20 hrs hydraulic retention time (HRT) and influent COD ranging between 2000-8000 mgL^{-1} showed the efficiency of organic content reduction in the reactor ranged from 75% to 93%. The reactor filled with standard pall rings made of polypropylene with an effective surface area of 206 m^2m^{-3} performed best in comparison to the reactor filled with cut polyethylene pipe 134 m^2m^{-3} and reactor filled with PVC packing (50 m^2m^{-3}). There was 2-7% decrease in efficiency with PE while it was 10-16% in case of PVC when compared to standard pall rings. Distillery waste tertiary treatment in a laboratory stabilization pond was evaluated. Effluents from a combined laboratory-scale anaerobic filter-aerobic trickling filter process were used as the influent. The effect of the hydraulic retention time (HRT) in the range of 5 - 30 days was evaluated at influent total chemical oxygen demand (TCOD) in the range of 271 - 5286 mgL^{-1} (BOD_5 in the range of 66-1212 mgL^{-1}). TCOD and BOD_5 removals of up to 54% and 74%, respectively, were obtained for the most concentrated influent used at 30 days of HRT. The effect of the HRT on TCOD and BOD_5 removal efficiencies followed an exponential relationship. An empirical equation adequately described the effect of the organic surface-loading rate on the process efficiency.

The pond acted as an intermediate flow pattern between completely mixed flow and plug-flow. It was found that a second-order model fit in well with the experimental results of organic matter removal rate and effluent substrate concentration. The second-order kinetic constant was found to be KS = 0.23 mg BOD_5/mg TSS day. The proposed model accurately predicted the behaviour of the pond showing deviations between the experimental and theoretical values of effluent substrate concentrations equal to or lower than 11% [55]. Sharma *et al.* [56] worked on the effect of nutrients supplementation on anaerobic sludge development and activity for treating distillery effluent. Startup of laboratory anaerobic reactors and treatment efficiency were investigated by supplementing the distillery effluent feed with macronutrients (Ca, P) and micronutrients (Ni, Fe and Co) under mesophilic conditions. Calcium and Phosphate were detrimental to the treatment efficiency and sludge granulation. Traces of salt of Iron, nickel and cobalt individually and in combinations improved the COD removal efficiency and sludge granulation process.

Driessen *et al.* [57] studied on the anaerobic treatment of distillery effluent with the UASB process Special attention was given to the treatment of effluents from sugar cane based distilleries with very high COD concentration of 60000 to 160000 mg COD. Despite expected toxicity problems. On the other hand, two identical UASB reactors operated in parallel as duplicates for 327 days for the treatment of malt whisky pot ale and achieved COD reductions of up to 90% for influent concentrations of 3526-52126 mgL^{-1}. When the OLRs of 15 kg m^{-3} day and above were used, the COD removal efficiency dropped to less than 20%. A mesophilic two stage system consisting of an anaerobic filter (AF) and an UASB reactor was found suitable for anaerobic digestion of distillery waste, enabling better conditions for the methanogenic phase [58]. Immobilization of bacteria in biofilm and on bioflocs is a crucial step in anaerobic degradation because of advantages such as higher activities, higher COD removal percent at short hydraulic retention times and better tolerance to disturbances such as toxic and organic shock loadings. At the same time there were certain disadvantages as well because in addition to some readily biodegradable matter, vinasses contain compounds like phenols, which were toxic to bacteria and inhibit the digestion. Also, due to seasonal nature of many of these industries and the absence of microorganisms in vinasses capable of carrying out anaerobic digestion, long incubation periods are required for the start-up stage. Besides, other operational problems in anaerobic digestion such as low growth rate of anaerobic bacteria and the loss of biomass in systems with high hydraulic rates frequently does not achieve a satisfactory purification of vinasses [59]. The formation of H_2S in anaerobic reactors is the result of the reduction of oxidized sulphur compounds. Methanogenic bacteria can tolerate sulphide concentration up to 1000 mgL^{-1} total sulphide. A complete loss of methane production occurred at 200 mgL^{-1} of un-ionized H_2S during digestion of flocculent sludge. Anaerobic contact process incorporating an ultra filtration (UF) unit was used to treat distillery wastewater characterized by high and low carbon to nitrogen concentrations. This treatment system showed methane yield of up to 0.6 m^3 kg^{-1} VS and removed up to 80% of the volatile acids [60]. Two-phase anaerobic digestion of alcohol stillage proved to be superior to the single-phase process in terms of substrate loading rate and methane yield, without affecting the treatment performance [61]. While maintaining BOD and COD reduction of 85% and 65% respectively, the two-phase achieved methane yield three times that of single-phase system.

Harada *et al.* [62] reported 39-67% COD removal, with a corresponding BOD removal of over 80%. The results suggested that the wastewater contained high concentration of refractile compounds; this, in turn, affected the microbial population in the sludge granules. Generally, the predominant genera of methanogens in granular sludge are *Methanobacterium, Methano-*

brevibacter, Methanothrix and *Methanosarcina* [63]; however, the predominance of *Methanothrix* in granular sludge is most essential for the establishment of a high performance UASB process. In this study, abundance of *Methanosarcina* sp. was observed whereas *Methanothrix* sp. was present to a lesser extent thereby indicating that the latter was more sensitive to refractile compounds. Influence of reactor configuration on fermentative hydrogen (H_2) production and substrate degradation was evaluated employing anaerobic mixed consortia. Reactors were operated at acidophilic (pH 6.0) condition employing designed synthetic wastewater as substrate at an organic loading rate of 3.4 Kg COD/m^3/day with a retention time of 24 hrs at 28 ± 2 °C. Experimental data enumerated the influence of reactor configuration on both H_2 production and wastewater treatment. Biofilm reactor (28.98 mmol H_2/day; 1.25 Kg COD/m^3/day) showed relatively efficient performance over the corresponding suspended growth configuration (20.93 mmol H_2/day; 1.08 Kg COD/m^3/day). Specific H_2 yields of 6.96 mmol H_2/g-COD_L-day (19.32 mmol H_2/g-COD_R/day) and 5.03 mmol H_2/g-COD_L/day (16.10 mmol H_2/g-COD_R/day) were observed during stabilized phase of operation of biofilm and suspended growth reactors respectively. Higher concentration of VFA generation was observed in the biofilm reactor. Both the configurations recorded higher acetate concentration over other soluble metabolites indicating the dominance of acid-forming metabolic pathway during the H_2 production process [64]. A lab-scale anaerobic hybrid (combining sludge blanket and filter) reactor was operated in a Continuous mode to study anaerobic biodegradation of distillery effluent. The study demonstrated that at optimum hydraulic retention time (HRT), 5 days and organic loading rate (OLR), 8.7 kg COD / m^3 d, the COD removal efficiency of the reactor was 79%. The anaerobic reduction of sulfate increases sulfide concentration, which inhibited the metabolism of methanogens and reduced the performance of the reactors. The kinetics of biomass growth *i.e.* yield coefficient (Y, 0.0532) and decay coefficient (Kd, 0.0041 d^{-1}) was obtained using Lawrence and McCarty model. However, this model failed in determining the kinetics of substrate utilization. Bhatia *et al.* [65] model having inbuilt provision of process inhibition described the kinetics of substrate utilization, *i.e.* maximum rate of substrate utilization and inhibition coefficient values. Modeling of the reactor demonstrated that Parkin *et al.* [66], and Bhatia *et al.* [65] models, both, could be used to predict the effluent substrate concentration. However, Parkin *et al.* [66] model predicts effluent COD more precisely (within ± 2 %) than Bhatia *et al.* [65] model (within ± 5 - 20 %) of the experimental value. Karhadkar *et al.* [67] model predicted biogas yield within ± 5 % of the experimental value. Besides these anaerobic treatments another treatment i.e. aerobic treatment is required for the safe disposal of effluent in the environment.

4.2. Aerobic treatment

Anaerobically treated distillery wastewater still contains high concentrations of organic pollutants and then cannot be discharged directly. The partially treated effluent has high BOD, COD and suspended solids. It can reduce the availability of essential mineral nutrients by trapping them into immobilized organic forms, and may produce phytotoxic substances during decomposition. Stringent regulations on discharge of colored effluent impede direct discharge of anaerobically treated effluent [52]. Therefore, aerobic treatment of sugarcane molasses wastewater has been mainly attempted for the decolorization of the major colorant, melanoidins, and for reduction of the COD and BOD. A large number of microorganisms such as bacteria (pure and mixed culture), cyanobacteria, yeast and fungi have been isolated in recent years and are capable of degrading melanoidins and thus decolorising the molasses wastewater. The aerobic methods have been described below.

4.2.1. Bacterial treatment

Numerous bacteria capable of melanoidins decolorization have been reported. Spent wash is pollution intensive wastewater generated by distilleries. Its dark brown color is due to recalcitrant melanoidin pigments. Bacterial cultures were screened for their ability to degrade these pigments.

Recently Lactic acid bacteria (Lactobacillus coryniformis, Lactobacillus sakei, Lactobacillus plantarum, Weisella soli, Pediococcus parvulus, Pediococcus pentosaceus) were used to decolorise the melanoidins. The isolate Lactobacillus plantarum exhibit 44% decourisation of melanoidins [68]. An effort has also been made by the Kryzwonos [69] using the consortia of genus Bacilli to decolorize the melanoidins. Two mixed bacterial cultures of the genus Bacillus (C1 and C2) were tested for color removal ability. Similarly Yadav and Chandra [70] developed the consortia Proteus mirabilis (IITRM5; FJ581028), Bacillus sp. (IITRM7; FJ581030), Raoultella planticola (IITRM15; GU329705) and Enterobacter sakazakii (IITRM16, FJ581031) in the ratio of 4:3:2:1. This consortia was responsible for the 75% decolorization of melanoidins within 10 days (Figure 3).

The isolate *Alcaligenes faecalis* strain SAG_5 showed 72.6 % decolorization of melanoidins at optimum pH (7.5) and temperature (37 °C) on 5th day of cultivation. The toxicity evaluation with mung bean (*Vigna radiata*) revealed that the raw distillery effluent was environmentally highly toxic as compared to biologically treated distillery effluent, which indicated that the effluent after bacterial treatment is environmentally safe [71]. The degradation of synthetic and natural melanoidins was studied by using the axenic and mixed bacterial consortium (*Bacillus licheniformis* (RNBS1), *Bacillus* sp. (RNBS3) and *Alcaligenes* sp. (RNBS4). Results have revealed that the mixed consortium was more effective compared to axenic culture decolorising 73.7% and 69.8%% synthetic and natural melanoidins whereas axenic cultures RNBS1, RNBS3 and RNBS4 decolorized 65.88%, 62.5% and 66.1% synthetic and 52.6%, 48.9% and 59.6% natural melanoidins, respectively. The HPLC analysis of degraded samples has shown reduction in peak areas compared to controls, suggesting that decrease in color intensity might be largely attributed to the degradation of melanoidins by isolated bacteria [72]. Tondee and Sirianuntapiboon [73] isolated *Lactobacillus plantarum* No. PV71-1861 from pickle samples in Thailand. The strain showed the highest melanoidin pigment (MP) decolorization yield of 68.12% with MP solution containing glucose 2%, yeast extract 0.4%, KH_2PO_4 0.1%, $MgSO_47H_2O$ 0.05% and initial pH 6 under static condition at 30 °C within 7 days. But, it showed low growth and MP decolorization yields under aerobic conditions. Gel filtration chromatograms of the MP solutions showed that the small molecular weight fraction of MP solution was decolorized by the strain when the large molecular weight fraction still remained in the effluent. For application, the strain could apply to treat anaerobic treated-molasses wastewater (T-MWW) with high removal efficiency. The highest MP removal efficiencies and growth yield of 76.6% and

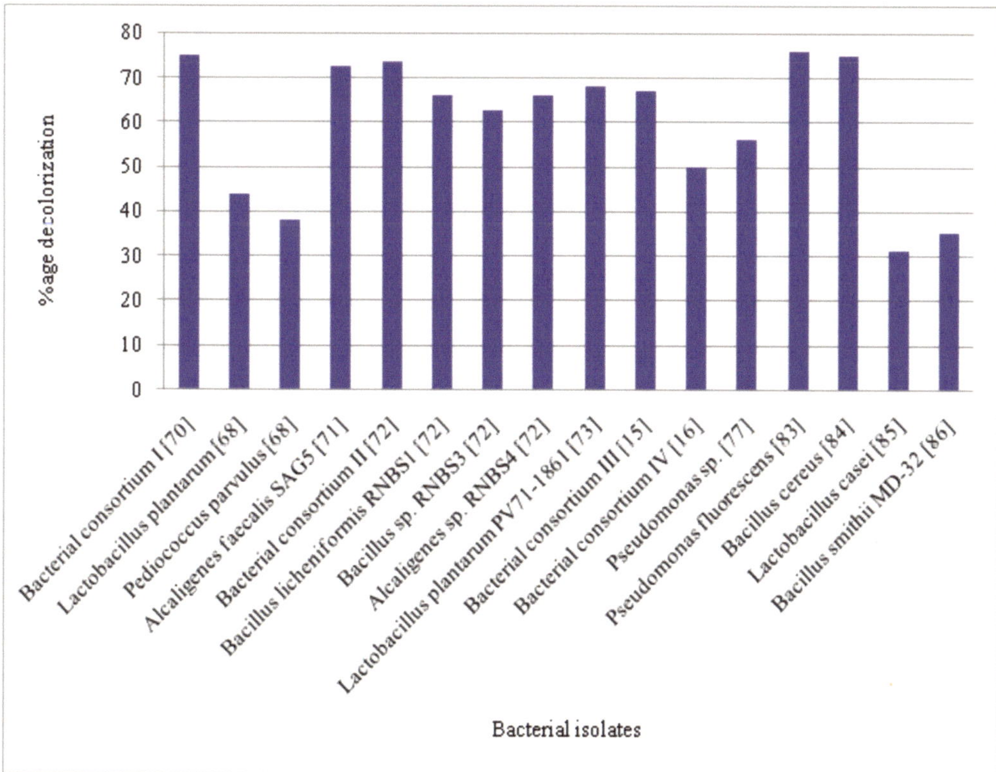

Figure 3. Melanoidin decolorization activity by different bacterial cultures. Bacterial consortium I: *Proteus mirabilis* (IITRM5; FJ581028), *Bacillus* sp. (IITRM7; FJ581030), *Raoultella planticola* (IITRM15; GU329705) and *Enterobacter sakazakii* (IITRM16, FJ581031) in the ratio of 4:3:2:1[70]; Bacterial consortium II: *Bacillus licheniformis* (RNBS1), *Bacillus* sp. (RNBS3) and *Alcaligenes* sp. (RNBS4) [72]; Bacterial consortium III: *Pseudomonas aeruginosa* PAO1, *Stenotrophomonas maltophila* and *Proteus mirabilis* [15]; Bacterial consortium IV: *Bacillus thuringiensis* (MTCC 4714), *Bacillus brevis* (MTCC 4716) and *Bacillus* sp. (MTCC 6506) [16].

2.6 mg/ml, respectively, were observed with the T-MWW within 7 days of culture, and the effluent pH of the system was decreased to lower than 4.0 after 2-3 days operation.

The identification of culturable bacteria by 16S rDNA based approach showed that the consortium composed of *Klebsiella oxytoca, Serratia mercescens, Citrobacter* sp. and unknown bacterium decolorise the synthetic melanoidins. In the context of academic study, prevention on the difficulties of providing effluent as well as its variations in compositions, several synthetic media prepared with respect to color and COD contents based on analysis of molasses wastewater, *i.e.*, Viandox sauce (13.5%, v/v), caramel (30%, w/v), beet molasses wastewater (41.5%, v/v) and sugarcane molasses wastewater (20%, v/v) were used for decolorization using consortium with color removal 9.5%, 1.13%, 8.02% and 17.5%, respectively, within 2 days. However, Viandox sauce was retained for further study. The effect of initial pH and Viandox concentration on decolorization and growth of bacterial consortium were further determined. The highest decolorization of 18.3% was achieved at pH 4 after 2 day of incubation [74]. Mohana *et al.* [15] isolated the microorganisms that were capable of decolorizing anaerobically treated distillery spent wash. A bacterial consortium DMC comprising of three bacterial cultures was selected on the basis of rapid effluent decolorization and degradation, which exhibited 67 ± 2% decolorization within 24 hrs and 51 ± 2% chemical oxygen demand reduction within 72 hrs when incubated at 37 °C under static condition in effluent supplemented with glucose 0.5%, KH_2PO_4 0.1%, KCl 0.05% and $MgSO_4.7H_2O$ 0.05%. Addition of organic or inorganic nitrogen sources did not support decolorization. The cultures were identified as *Pseudomonas aeruginosa* PAO1, *Stenotrophomonas maltophila* and *Proteus mirabilis* by the 16S rDNA analysis. Kumar and Chandra [16] studied on the decolorization of synthetic melanoidins (*i.e.*, GGA, GAA, SGA and SAA) by three *Bacillus* isolates *Bacillus thuringiensis* (MTCC 4714), *Bacillus brevis* (MTCC 4716) and *Bacillus sp.* (MTCC 6506). Significant reduction in the values of physicochemical parameters was noticed along with the decolorization of all four melanoidins (10%, v/v). *B. thuringiensis* (MTCC 4714) caused maximum decolorization followed by *B. brevis* (MTCC4716) and *Bacillus* sp. (MTCC6506). A mixed culture comprised of these three strains was capable of decolorizing all four melanoidins. The medium that contained glucose as sole carbon source showed 15% more decolorization than that containing both carbon and nitrogen sources. Melanoidin SGA was maximally decolorized (50%) while melanoidins GAA was decolorized least in the presence of glucose as a sole energy source. Acetogenic bacteria are capable of oxidative decomposition of melanoidins. Adikane *et al.* [75] worked decolorization of molasses spent wash (MSW) in absence of any additional carbon or nitrogen source using soil as inoculum. A decolorization of 69% was obtained using 10% (w/v) soil and 12.5% (v/v) MSW after 7 days incubation. Optimized parameters including days 6 days, pH 6, MSW 12.5% and soil concentration 40%, were obtained for maximum decolorization. A decolorization of 81% was achieved using 10% soil and 12.5% MSW after 18 days incubation in absence of any media supplement. Nearly 12% reduction in decolorization activity of the soil sample was observed over a period of 12 months when stored at 6 °C. It could be concluded that the decolorization of MSW might be achieved using soil as inoculum without addition of chemical amendments. On the other hand, *Bacillus megaterium* SW3 and *Bacillus subtilis* SW8 were used for decolorization and bioremediation of DMSW. Both bacterial isolates were grown well in 7.5% (v/v) diluted digested spent wash supplemented with glucose and urea as a readily available carbon and nitrogen source. Optimum condition for growth and DMSW decolorization were at pH 7.2, temperature 37°C, glucose 5% (w/v), and urea 2% (w/v) in a minimal salt medium. The maximum decolorization (51%, 52.5% and 57%) and COD reduction (53%, 54.5% and 59%) was found after 5 days of incubation under the optimized condition were achieved for cultures SW3, SW8 and consortia (SW3 + SW8) respectively [76].

Chavan *et al.* [77] isolated a *Pseudomonas* sp. and studied for degradation of recalcitrant melanoidin pigment. The optimum conditions for decolorization were pH 6.8 -7.2, temperature 30-35 ºC and glucose 0.4% (w/v). The maximum decolorization up to 56% and 63% reduction in COD of the spent wash could be achieved after 72 hrs by the microbial treatment. Spectrophotometric and HPLC analysis of treated effluent confirmed biodegradation of melanoidin pigments by the isolate. This approach could be used to develop a cost-effective, eco-friendly biotechnology package for the bioremediation of spent wash before its disposal.

Ghosh *et al.* [78] worked on enrichment and identification of bacteria capable of reducing COD of anaerobically treated molasses spent wash. The isolates were grouped into six haplotypes by amplified ribosomal DNA restriction analysis (ARDRA) and BOX-PCR. They showed maximum similarity to six genera *viz. Pseudomonas, Enterobacter, Stenotrophomona, Aeromonas, Acinetobacter* and *Klebsiella.* The extent of COD (44%) reduced collectively by the six strains was equal to that reduced individually by *Aeromonas, Acinetobacter, Pseudomonas* and *Enterobacter.* With spent wash as sole carbon source, the COD reducing strains grew faster at 37 °C than 30 °C. Sirianuntapiboon *et al.* [79] isolated 170 strains of acetogenic bacteria, a strain No.BP103 showed the highest decolorization yield when cultivated at 30 °C for 5 days in molasses pigments medium containing glucose 3.0%, yeast extract 0.5%, KH_2PO_4 0.1% and $MgSO_4.7H_2O$ 0.05% and the pH adjusted to 6.0. In addition this strain could decolorise 32.3% and 73.5% of molasses pigments in stillage and anaerobically treated molasses wastewater, both supplementing glucose 3.0% yeast extract 0.5%, KH_2PO_4 0.1% and $MgSO4.7H_2O$ 0.05%. However without nutrients supplement, the decolorization yields were 9.75% and 44.3% respectively. In a replacement culture system involving six time replacement (30days), the strain No. BP103 Showed a constant decolorization yield of 72.0% to 84.0% and caused decreases of biological oxidation demand and chemical oxygen demand values of approximately 58.5% to 82.2% and 35.5% to 71.2%, respectively. Under a periodical feeding system, the decolorization yield was 30.0 to 45.0%. Cibis *et al.* [80] achieved biodegradation of potato slops (distillation residue) by a mixed population of bacteria under thermophilic conditions up to 60 °C. A COD removal of 77% was achieved under non-optimal conditions.

Savant *et al.* [81] isolated a novel acid-tolerant, hydrogenotrophic methanogenic isolate ATM^T, using enrichment culture technique at pH 5 using slurry from an acidogenic digester running on alcohol distillery waste. The original pH of the slurry was 5.7 and the volatile fatty acid concentration was 9000 ppm. Cells of isolate ATM^T were Gram-positive, non-motile and 0.3-0.5 μm in size. They did not form spores. The isolate could grow in the pH range 5.0-7.5, with maximum growth at pH 6.0. The optimum temperature for growth was 35 °C. Morphological and biochemical characteristics of the isolate, together with the 16S rDNA sequence analysis, clearly revealed that the isolate could be accommodated within any of the existing species of the genus *Methanobrevibacter.* Therefore, it is proposed that a novel species of the genus *Methanobrevibacter* should be created for this isolate, *Methanobrevibacter acididurans* sp. nov., and the type strain is strain ATM^T. Ghosh *et al.* [82] studied on the application of two bacterial strains, *Pseudomonas putida* U and *Aeromonas* strain Ema, in a two-stage column bioreactor to reduce COD and the color of anaerobic treated spent wash under aerobic conditions. *P. putida* reduced the COD and color by 44.4% and 60%.

Dahiya *et al.* [83] worked on decolorization of molasses wastewater by cells of *Pseudomonas fluorescens*. Decolorization was 76% under nonsterile conditions in four days at 30 °C. Immobilised cells could be reused for decolorization activity. However, in subsequent cycles, this was found to decrease from 76% to 50% and from 50% to 24%. Decolorization activity was regenerated from 30% to 45% by recultivating the immobilised cells in a fresh growth medium. Cellulose carrier coated with collagen was found to be most effective carrier, which produced the highest decolorization activity of 94% in a day process. This carrier could be reused with

50% of the decolorization activity retained until the seventh day. Jain *et al.* [84] studied on degradation of anaerobically digested distillery wastewater by three bacterial strains, *viz. Xanthomonas fragariae, Bacillus megaterium* and *Bacillus cereus* in free and immobilized form, isolated from the activated sludge of a distillery wastewater treatment plant. The removal of COD and color with all the three strains increased with time up to 48 hrs and only marginal increase in COD and color removal efficiency was observed beyond this period up to 72 hrs. After this period, removal efficiency remained fairly constant up to 120 hrs. The maximum COD and color removal efficiency varied from 66 to 81 % and 65 to 75 % respectively for both free and immobilized cells of all the three strains. The strain *Bacillus cereus* showed the maximum efficiency of COD (81 %) and color (75 %) removal out of the three strains.

Masaharu *et al.* [85] studied on the screening of acid-forming facultative anaerobes for their decolorizing ability with the distilled wastewater (DWDA) produced during the alcoholic fermentation of cane molasses was carried out. On the primary screening, 113 strains, which were isolated from enrichment cultures of various plant and animal samples or soil samples showed decolorizing activity on modified GYP medium containing 10% (v/v) distilled wastewater. Strain SM-3 was selected as the best strain. This strain decolorized 31.4% of molasses pigment at a concentration of 10% (v/v) distilled wastewater supplemented with 1% glucose within 5 days at 30 °C under semi-anaerobic conditions. The main DWDA was found in the cell-free extract and cell membrane fractions. The strain was identified as *Lactobacillus casei*. Kambe *et al.* [86] screened out various microorganisms for their ability to decolorise molasses wastewater under thermophilic and anaerobic conditions. Strain MD-32, which was newly isolated from a soil sample, was selected as the best strain. From taxonomical studies, the strain was concluded to belong to the genus *Bacillus*, most closely resembling *B. smithii*. The strain decolorized 35.5% of molasses pigment within 20 days at 55 °C under anaerobic conditions but no decolorization activity was observed when it was cultivated aerobically.

At all the concentration tested molasses pigment was effectively decolorized by MD-32, with decolorization yields of around 15% within 2 days. The molecular weight distribution as determined by gel filtration chromatography revealed that the decolorization of molasses pigment by the isolated strain is accompanied by a decrease in not only small molecules but also large ones. Sirianuntapiboon [87] isolated acid forming bacteria from vegetables, fruits and fermented food samples. They were primary screened to select the strains which had decolorization activity. The results showed that 50 isolates had decolorization activity on solid medium (clear zone). Among these, the strains No. BP103 and No. 13A gave the highest decolorization activity in the liquid medium which contained molasses pigments. Fructose was the most suitable carbon source for both strains for decolorizing molasses pigments. The optimal concentration of fructose in the media for highest decolorizing activity was 2.0%. For the effects of nitrogen source on decolorization activity, the organic nitrogen compounds were the best nitrogen sources for both strains. The decolorization activity of the strain No. 13A and No. BP103 were 80.5% and 82.0%, respectively when the nitrogen source was yeast extract. At the optimal conditions, the decolorizing activity of the strains No. 13A and BP103 were 90.5% and 96.7%, respectively. Strain No. 13A and No. BP103 were identified as *Acetobacter acetii*. Ohmomo *et al.* [88] performed screening of facultative anaerobes with melanoidin decolorizing activity (MDA). Strains were isolated from stored wastewater of an alcohol fermentation involving molasses. One of them, strain W-NS, showed high and stable MDA, and was identical with *Lactobacillus hilgardii*. The decolorization yield of this strain under the optimal conditions was improved to 40% by immobilization of cells within Ca- alginate gel.

4.2.2. Fungal treatment

Increasing attention has been directed towards utilizing microbial activity for decolorization of molasses wastewater. Several reports have indicated that some fungi in particular have such a potential [89]. One of the most studied fungus having ability to degrade and decolorise distillery effluent is *Aspergillus* such as *Aspergillus fumigatus* G-2-6, *Aspergillus niger*, *Aspergillus niveus*, *Aspergillus fumigatus* UB260 brought about an average of 69-75% decolorization along with 70-90% COD reduction [22, 90-92]. Previous studies showed that the white-rot fungus *Phanerochaete chrysosporium* could remove color and total phenols from the sugar refinery effluent [93]. Ohmomo *et al.* [94] reported the continuous decolorization of molasses waste-water in a bubbling column reactor with *Coriolus versicolor* immobilized within Ca-alginate gel. Later, Ohmomo *et al.* [95] used autoclaved mycelium of *Aspergilus oryzae* Y-2-32 that adsorbed lower weight fractions of melanoidins and degree of adsorption was influenced by the kind of sugars used for cultivation. The wine distilleries produce large volume of waste-water having phenolic compounds, which give a high inhibitory and antibacterial activity to this wastewater, thus slowing down the anaerobic digestion process. Partial elimination of these phenolics compounds was obtained by using *Geotrichum candidum* [96]. *Rhizoctonia* sp. D90 decolorized molasses melanoidins medium and a synthetic melanoidin medium by 87.5% and 84.5%, respectively, under experimental conditions [97] (Figure 4). Murata *et al.* [98] isolated a fungus *Streptomyces werraensis* TT 14 from soil that decolorized the model melanoi-din, the decolorization rate being 64% in the optimal medium of pH - 5.5 and 45% in a synthetic medium. There was virtually no difference in the UV-VIS absorption spectra of the microbially treated melanoidin and control. The peaks of the gel permeation chromatogram for the treated melanoidin and for the control showed the same retention times, but lower molecular weight compounds increased in the decolorized melanoidin. Kida *et al.* [99] *Aspergillus awamori* var. *kawachi* has been used for production of single cell protein from Japanese distillery wastewater after aerobic cultivation.

Fujita *et al.* [100] also considered the decolorization of melanoidin present in an effluent with *Coriolus hirsutus*. The aim of this study was to find a suitable industrial biological treatment process for efficient decolorization of sugar refinery effluent. White-rot fungi have been reported to be capable of decolorizing melanoidin containing wastewaters. Previous studies showed that the white-rot fungus *Phanerochaete chrysosporium* can remove color and total phenols from the sugar refinery effluent [93]. Several studies have been reported for decolorizing the distillery effluent using fungi. Fumi *et al.* [101] worked on the optimisation of long term activated sludge treatment of winery wastewater. The results obtained in work carried out in order to verify the overall efficiency of full-scale, long-term, activated- sludge treatment of winery wastewater. The analytical data showed the high removal of COD during the whole experimentation period and with various working parameters. Sludge production was lower than that produced by ordinary

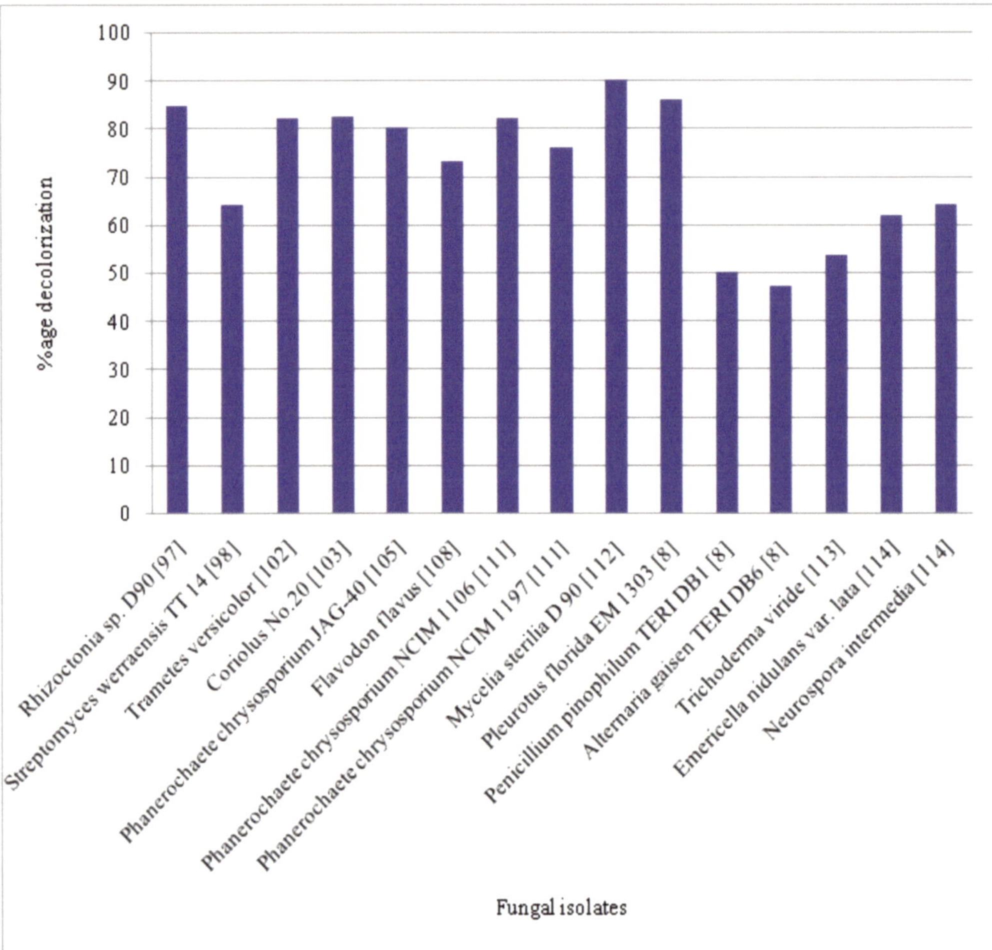

Figure 4. Melanoidin decolorization activity by different fungi.

activated-sludge plants. Miranda *et al.* [90] worked on color elimination from molasses wastewater by *Aspergillus niger*. Color elimination by *Aspergillus niger* from wastewater was studied. The influences of the nutrient concentrations, initial pH and carbon source on this color elimination were analyzed. During the batch process, through all experiments the maximal color elimination was attained after 3 or 4 days in the culture. Batch processes showed a maximal color elimination of 69% when $MgSO_4$, KH_2PO_4, NH_4NO_3 and a carbon source were added to the wastewater. The continuous process, with the same nutrient concentrations, showed less color removal and the decolorization activity was maintained for only 4 days. Benito *et al.* [102] studied on decolorization of wastewater from an alcoholic fermentation process with *Trametes versicolor*. Tests varying the concentrations of carbon source, nutrients, initial pH and mycelia were carried out in order to analyze their influence on the percentage decolorization, COD removal and decrease of ammonium content of the wastewater. In this way, 82% color elimination, 77% COD removal and 36% N-NH: decrease was attained.

Another fungus, *Coriolus* No.20, belonging to the class basidiomycetes, can remove color substances from molasses wastewater (MWW). In the case of treatment of MWW solution, the decrease was 82.5% in darkness under optimal conditions. This decolorization was through the adsorption of melanoidin to the mycelium and the yield of adsorption was 9.67 and 12.2 mg g^{-1} of mycelium as dry basic in the case of autoclaved mycelium and resting mycelium, respectively. The resting mycelium showed the highest adsorption yield (16.32 mg g^{-1} of mycelium) when the melanoidins solution was supplemented with 20% glucose solution. In the fed-batch system, the strain showed a constant decolorization yield of 75% during four times replacements (32 days) in both untreated and treated MWW solutions. In the continuous feed system for decolorization of untreated MWW solution, l0% of fresh untreated MWW solution was added every day after seven days of cultivation. The decolorization rate was constant (37.5%) during 12 days of operation. In the continuous decolorization of concentrated untreated MWW solution, the strain showed a constant decolorization yield (about 70%) during three times feeding of fresh concentrated untreated MWW solution. But I0 ml of 20% glucose solution had to be added after 13 days cultivation to keep the reducing sugar concentration in the culture broth at more than 1.0% [103].

Miyata *et al.* [26] worked on Manganese-Independent and Dependent decolorization of melanoidin using fungi. *Coriolus hirsutus* pellets were grown in a melanoidin-containing medium produced extracellular H_2O_2 up to 43 μm. However, nonenzymatic decolorization of melanoidin by H_2O_2, which has been previously reported, seemed very little in the fungal culture. The culture fluid contained two extracellular peroxidases, a manganese-independent peroxidase (MIP) and manganese peroxidase (MnP). Since both MIP and MnP showed melanoidin-decolorizing activities in the presence of H_2O_2, it was resulted that melanoidin decolorization in the *C. hirsutus* culture involved the production of extracellular H_2O_2 and the peroxidases. Miyata *et al.* [104] studied on the combined use of activated sludge and the fungus *Coriolus hirsutus*. A white rot fungus, *Coriolus hirsutus*, exhibited a strong ability to decolorise melanoidin in cultures not supplemented with nitrogenous nutrients. Addition of peptone to the cultures lowered the ability of the fungus to decolorise melanoidin, but that of inorganic nitrogens, ammonium and nitrate did not bring about any marked reduction in the ability. It suggested an inhibitory effect of organic nitrogen on melanoidin decolorization. Therefore, for enhancing the decolorization of melanoidin in wastewaters by the fungus, activated sludge pretreatment of the wastewaters was expected to be effective, *i.e.* activated sludge is capable of converting available organic N into inorganic N. To confirm this, waste sludge heat treatment liquor (HTL), wastewater from a sewage treatment plant, was pretreated with activated sludge. In practice, pretreatment of HTL under appropriate conditions accelerated the fungal decolorization of HTL. In the pretreated HTL, the fungus was shown to produce a high level of manganese-independent peroxidase (MIP). Addition of Mn(I1) to the pretreated HTL caused a further increase in the decolorization efficiency of the fungus and a marked increase in the manganese peroxidase (MnP) activity. Consequently, the increases in MIP and MnP activities were considered to play an important role in the enhanced ability of *C. hirsutus* to decolorise HTL.

Dahiya *et al.* [105] worked on decolorization of synthetic and spentwash melanoidins using the white-rot fungus *Phanerochaete chrysosporium* JAG-40. This fungus was isolated from the soil samples saturated with spilled molasses collected from a sugar mill. This isolate decolorized synthetic and natural melanoidins present in spentwash in liquid fermentation; up to 80% in 6 days at 30 °C under aerobic conditions. A large inoculum size stimulated fungal biomass production, but this gave less decolorization of pigment; 5% w/v (dry weight) mycelial suspension was found optimum for maximum decolorization in melanoidin medium supplemented with glucose and peptone. Gel-filtration chromatography showed that larger molecular weight fractions of melanoidins were decolorized more rapidly than small molecular weight fractions. Malandra *et al.* [106] studied the microorganisms associated with a rotating biological contactor (RBC) treating winery wastewater. One of the yeast isolates was able to reduce the COD of synthetic wastewater by 95% and 46% within 24 hrs under aerated and non aerated conditions, respectively. Kwak *et al.* [107] worked on effect of reaction pH on the photodegradation of model melanoidins. The wastewaters of molasses-based alcohol distilleries contain brown colored melanoidin pigments that are one of the major pollutants.

Ragukumar *et al.* [108] reported decolorization of such intensely brown colored molasses spent wash (MSW) by *Flavodon flavus,* a white-rot basidiomycete fungus isolated from a marine habitat. They attempted to improve the process of decolorization of MSW by this fungus by immobilization. Polyurethane foam-immobilized-fungus decolorized 10% diluted MSW by 60% and 73% by day 5 and 7 respectively. The immobilized fungus could be effectively used for a minimum of 3 cycles repeatedly to decolorise MSW. Besides decolorization, the fungus also removed the toxicity of MSW. Raghukumar and Rivonkar, [109] studied on the decolorization of molasses spent wash by the white rot fungus *Flavodon flavus* isolated from decomposting leaves of a sea grass, decolorized pigment in molasses spentwash (MSW) by 80% after 8 days of incubation, when used at concentrations of 10% and 50%. Decolorizing activity was also present in the media prepared with half strength sea water. Decolorizing activity was also seen in low nitrogen medium, nutrient rich medium and in sugar cane bagasse medium. The percentage decolorization of MSW was highest when glucose or sucrose was used as a carbon source in the low nitrogen medium

On the other hand, *Phanerochaete chrysosporium* was used for the vinasse degradation under two different growth conditions. Vinasse was treated by *P. chrysosporium* in a liquid inoculum form, during 32 days at room temperature (approximately 25 °C) and at 39 °C. The chemical oxygen demand (COD), total phenol concentration and color removal were measured and there was a decrease in COD, phenolic concentration and color of 47.4%, 54.7% and 45.1% respectively, at room temperature and a decrease in 54.2%, 59.4% and 56.8% respectively at 39 °C [110]. Thakkar *et al.* [111] studied the biocatalytic decolorization of molasses by *Phanerochaete chrysosporium*. Bioremediation potential of *Phanerochaete chrysosporium* strains NCIM 1073, NCIM 1106 and NCIM 1197 to decolorise molasses in solid and liquid molasses media was studied. Strains varied in the pattern of molasses decolorization on solid medium by giant colony method. Under submerged cultivation conditions, strain NCIM 1073 did not decolorise molasses while, strains NCIM 1106 and NCIM 1197 could decolorise molasses up to 82% and 76%, respectively. Under stationary cultivation conditions, none of the strains could decolorise molasses.

Srianuntapiboon *et al.* [112] worked on three kinds of molasses wastewater collected from the stillage of an alcohol factory, and their chemical properties. *Mycelia sterilia* D 90 decolorized about 90% of the molasses pigment in 10 days and at the same time caused an about 80% decrease in the biological oxygen demand value when glucose 2.5%, $NaNO_3$ 0.2%, KH_2PO_4 0.1% and $MgSO_4.7H_2O$ 0.05% were added to the molasses wastewater from the stillage as nutrients. But without supplementation of the nutrients, the decolorization yield was only 17.5%. Furthermore, this strain showed a decolorization yield of about 70% in 11 days and caused a decrease in the biological oxygen demand value of about 90% in 15 days under non-sterile conditions. In the fed-batch system, this strain showed a constant decolorization yield of about 80% and caused a decrease in the biological oxygen demand value of about 70%.

Pant and Adholeya [8] isolated two fungal strains producing ligninolytic enzymes and having the potential to decolorise distillery effluent from the soil of a distillery effluent contaminated site. DNA was isolated from the pure cultures of these fungi and polymerase chain reaction (PCR) amplification of their internal transcribed spacer (ITS) region of nuclear ribosomal DNA was carried out. Further, the DNA was sequenced and the comparison of generated sequence with database led to their identification as *Penicillium pinophilum* TERI DB1 and *Alternaria gaisen* TERI DB6, respectively. These two isolates along with one isolate of *Pleurotus florida* EM 1303 were assessed for their ligninolytic enzyme activity in culture filtrate as well as after solid state fermentation on two substrates wheat straw and corncob powder. Ergosterol was measured to assess the growth of fungi on solid media. Both *P. pinophilum* TERI DB1 and *A. gaisen* TERI DB6 were found to produce laccase, manganese-dependent peroxidases (MnP) and lignin peroxidases (LiP). The immobilized fungal biomass was used for decolorization of the post biomethanated wastewater from the distillery. Reduction in color up to the magnitude of 86%, 50% and 47% was observed with *P. florida, P. pinophilum* and *A. gaisen* respectively. In another study, the fungi *Trichoderma viride* showed the highest decolorization yield (53.5%) when cultivated at 30 ºC for 7 days in medium contained the molasses which was diluted to 40 gL^{-1} in distilled water. The other *Trichoderma* species and *Penicillium* sp. also gave similar results of 40% - 45%. Decolorization yield was increased by adding peptone and yeast extract to the production medium except *Penicillium* sp. When the pH decreased below 5.0 after the incubation, the decolorization yield increased. Although reducing sugar in culture broth decreased with decreasing color intensity, there was no connection between protein utilization and decolorizing activity [113]. Two fungi (*Emericella nidulans* var. lata and *Neurospora intermedia)* have the capability to decolorise melanoidins. Maximum color was removed at pH 3, temperature 30 ºC, stirring 125 rpm, dextrose (0.05%) and sodium nitrate (0.025%) by both fungi. After optimization, there was two-fold increase in color removal from 38% to 62% (DF3) and 31% to 64% (DF4) indicating significance of Taguchi approach in decolorization of distillery mill effluent [114].

4.2.3. Treatment by other microorganisms

The treatment of anaerobically treated 10% distillery effluent using the microalga *Chlorella vulgaris* followed by *Lemna minuscula* resulted in 52% reduction in color [115]. In another study, Kalavathi *et al.* [116] examined the degradation of 5% melanoidin by the marine cyanobacterium *Oscillatoria boryana* BDU 92181. The organism was found to release hydrogen peroxide, hydroxyl ions and molecular oxygen during photosynthesis resulting in 60% decolorization of distillery effluent. In addition, this study suggested that cyanobacteria could use melanoidin as a better nitrogen source than carbon. Further, cyanobacteria also excrete colloidal substances like lipopolysaccharides, proteins, polyhydroxybutyrate (PHB), polyhydroxy-alkanoates (PHA), *etc.* These compounds possess COO^- and ester sulphate (OSO^{3-}) groups that can form complexes with cationic sites thereby resulting in flocculation of organic matter in the effluent. It was observed that the strain *Oscillatoria* resulted in almost complete color removal (96%) whereas *Lyngbya* and *Synechocystis* were less effective resulting in 81 and 26% color reduction, respectively [117]. The consortium of the three strains showed a maximum decolorization of 98%. This was attributed to adsorption in the initial stages followed by degradation of organic compounds which dominated in the subsequent stages. Sirianuntapiboon *et al.* [31] worked on decolorization of molasses wastewater by *Citeromyces* sp. WR-43-6. 205 yeast strains isolated from Thai-fruit samples were screened, and isolated strain No. WR-43-6 showed the highest decolorization yield (68.91%) when cultivated at 30°C for 8 days in a molasses pigments solution containing glucose 2.0%, sodium nitrate 0.1% and KH_2PO_4 0.1%, the pH being adjusted to 6.0. This potent strain was identified as *Citeromyces* sp. and showed highest removal efficiencies on stillage from an alcohol factory (U-MWW). The color intensity, chemical oxygen demand (COD) and biochemical oxygen demand (BOD) removal efficiencies were 75%, almost 100 and 76%, respectively. In a periodical feeding system, *Citeromyces* sp. WR-43-6 showed an almost constant decolorization yield of 60-70% over 8 days feeding of 10% fresh medium. In a replacement culture system, *Citeromyces* sp. WR-43-6 also gave a constant decolorization yield (about 75%) during four times replacement. Yeast *Citeromyces* was used for treating MWW and high and stable removal efficiencies in both color intensity and organic matter were obtained.

Moriya *et al.* [118] used two flocculant strains of yeast, *Hansenula fabianii* and *Hansenula anomala* for treatment of wastewater from beet molasses-spirits production and achieved 25.9% and 28.5% removal of TOC respectively from wastewater without dilution. Dilution of wastewater was not favourable for practical treatment of wastewater due to the longer treatment time and higher energy cost. Shojaosadati *et al.* [119] optimized the growth conditions for single cell protein (SCP) production and COD reduction by the use of *Hansenula* sp. in sugar beet stillage. They concluded that production of SCP from stillage is one of the most promising options. Besides, the yeast was also found to utilize lactate and acetate that are inhibitory to ethanol production. As a result, the treated effluent could be used as dilution water for fermentation thereby reducing the residual stillage volume by 70%.

5. Role of enzymes in effluent decolorization

The enzymatic treatment falls between the physicochemical and biological treatment processes. It has technological advantages and requires economical considerations to apply it on a large scale. It has some potential advantages over the conventional treatment. These includes: applicability to biorefractory compounds; operation either at high or low contaminant concentrations; operation over a wide range of pH, temperature and salinity; absence of shock loading effects; absence of delays associated with the acclimatization of biomass; reduction in the sludge volume and the ease and simplicity of controlling the process [120]. Recent research has focused on the development of enzymatic processes for the treatment of wastewaters, solid wastes, hazardous wastes and soils in recognition of these potential advantages [121]. A large number of enzymes (e.g. peroxidases, oxidoreductases, cellulolytic enzymes cyanidase, proteases, amylases, etc.) from a variety of different sources have been reported to play an important role in an array of waste treatment applications [122-127].

Although the enzymatic system related with decolorization of melanoidins is yet to be completely understood, it seems greatly connected with fungal ligninolytic mechanisms. The white-rot fungi have a complex enzymatic system which is extracellular and non-specific, and under nutrient-limiting conditions is capable of degrading lignolytic compounds, melanoidins, and polyaromatic compounds that cannot be degraded by other microorganisms [102]. A large number of enzymes from a variety of different plants and microorganisms have been reported to play an important role in an array of waste treatment applications. Several studies regarding degradation of melanoidins, humic acids and related compounds using basidiomycetes have also suggested a participation of at least one laccase enzyme in fungi belonging to *Trametes* genus. The role of enzymes other than laccase or peroxidases in the decolorization of melanoidins by *Trametes* strain was reported during the 1980s. Several reports claimed that intracellular sugar-oxidase- type enzymes (sorbose- oxidase or glucose-oxidase) had melanoidin-decolorizing activities. It was suggested that melanoidins were decolorized by the active oxygen (O_2; H_2O_2) produced by the reaction with sugar oxidases [128]. Decolorization by microbial methods includes the enzymatic breakdown of melanoidin and flocculation by microbially secreted substances. Ohmomo *et al.* [129] used *C. versicolor* Ps4a, which decolorized molasses wastewater 80% in darkness under optimum conditions. Decolorization activity involved two types of intracellular enzymes, sugar-dependent and sugar-independent. One of these enzymes required no sugar and oxygen for appearance of the activity and could decolorise MWW up to 20% in darkness and 11-17% of synthetic melanoidins. Thus, the participation of these H_2O_2 producing enzymes as a part of the complex enzymatic system for melanoidin degradation by fungi should be taken into account while designing any treatment strategy. One of the more complete enzymatic studies regarding melanoidin decolorization was reported by Miyata *et al.* [26]. Color removal of synthetic melanoidin by *C. hirsutus* involved the participation of peroxidases (MnP and MIP) and the extracellular H_2O_2 produced by glucose-oxidase, without disregard of a partial participation of fungal laccase. The manganese peroxidase (MnP) has also been reported in bacteria as extracellular enzyme for decolorization of melanoidin [72]. The involvement of MnP and laccase in white rot fungus for degradation of various biopolymers (lignin and tannin) has also been reported [130, 131]. But, the detail role of MnP and laccase in bacteria for decolorization of melanoidin has not been fully investigated.

Mansur *et al.* [132] obtained a maximum decolorization of around 60% on day 8 after inoculating with fungus *Trametes* sp. Here effluent was added at a final concentration of 20% (v/v) after 5 days of fungal growth, the time at which high levels of laccase activity were detected in the extracellular mycelium. The white-rot basidiomycete *T. versicolor* is an active degrader of humic acids as well as of melanoidins. A melanoidin mineralizing 47 kDa extracellular protein corresponding to the major mineralizing enzyme system from *T. versicolor* was isolated by Dehorter and Blondeau [133]. This Mn^{2+} dependent enzyme system required oxygen and was described to be as peroxidase. Uniform, small and spongy pellets of the fungus *T. versicolor* were used as inoculum for color removal using different nutrients such as ammonium nitrate, manganese phosphate, magnesium sulphate and potassium phosphate and also sucrose as carbon source [102]. Maximum color removal of 82% and 36% removal of $N\text{-}NH_4$ was obtained on using low sucrose concentration and KH_2PO_4 as the only nutrient. Some studies have identified the lignin degradation related enzymes participating in the melanoidin decolorization. Intracellular H_2O_2 producing sugar oxidases have been isolated from *Coriolus* strains. Also, *C. hirsustus* have been reported to produce enzymes that catalyze melanoidin decolorization directly without additions of sugar and O_2. Miyata *et al.* [26] used *C. hirsutus* pellets to decolorise a melanoidin-containing medium. It was elucidated that extracellular H_2O_2 and two extracellular peroxidases, a manganese-independent peroxidase (MIP) and manganese peroxidase (MnP) were involved in decolorization activity. Lee *et al.* [134] investigated the dye-decolorizing peroxidase by cultivating *Geotrichum candidum* Dec1 using molasses as a carbon source. Components in the molasses medium stimulated the production of decolorizing peroxidase but inhibited the decolorizing activity of the purified enzyme. It was found that the inhibitory effect of molasses can be eliminated at dilution ratios of more than 25. Recently D'souza *et al.* [135] reported 100% decolorization of 10% spent wash by a marine fungal isolate whose laccase production increased several folds in the presence of phenolic and non-phenolic inducers. A combined treatment technique consisting of enzyme catalyzed in situ transformation of pollutants followed by aerobic biological oxidation was investigated by Sangave and Pandit [121] for the treatment of alcohol distillery spent wash. It was suggested that enzymatic pretreatment of the distillery effluent leads to *in situ* formation of the hydrolysis products, which have different physical properties and are easier to assimilate than the parent pollutant molecules by the microorganisms, leading to faster initial rates of aerobic oxidation even at lower biomass levels. In another study, Sangave and Pandit [136] used irradiation and ultrasound combined with the use of an enzyme as pretreatment technique for treatment of distillery wastewater. The combination of the ultrasound and enzyme yielded the best COD removal efficiencies as compared to the processes when they were used as stand-alone treatment techniques. Enzymatic decolorization of molasses medium has also been tried using *P. chrysosporium* [111]. Under stationary cultivation conditions, none of the strains could decolorise molasses nor produce enzymes lignin peroxidase, manganese peroxidase and laccase. All of them could produce lignin peroxidase and manganese peroxidase when cultivated in flat bottom glass bottles under stationary cultivation conditions.

The MnP and laccase activity in the culture supernatant decreases with the increase in the time duration and further bacterial incubation showed gradual decrease of both enzyme activities. The enzyme activity has direct co-relation with the melanoidin decolorization. The initiation of MnP activity in culture supernatant starts at 48 h of bacterial growth and remains active up to 192 h, whereas, the laccase induction starts at 96 h and its activity decreased after 192 h. This resulted in the initial role of MnP in melanoidin degradation [70]. The melanoidin decolorization has also been reported by MnP activity in fungus [8, 108]. The enzyme laccase also helps in decolorizing melanoidin [72, 137].

The white-rot basidiomycetous fungus *F. flavus* produces extracelluar enzymes, manganese-dependent peroxidase (MNP), lignin peroxidase (LIP) and laccase which decolorizes a number of synthetic dyes like Poly R and remazol brilliant blue R etc. [138]. However, a negative correlation was found between MnP concentrations and decolorization of molasses spent wash (MSW) [139]. The levels of glucose oxidase activity also have correlation with decolorization of MSW. Raghukumar *et al.* [108] proposes that hydrogen peroxide produced as a result of glucose oxidase activity acts as a bleaching agent on MSW. Which has been further confirmed that in the absence of glucose the decolorization of MSW is lesser. With the addition of the starch to the medium, amylase activity stimulates and releasing reducing sugars. Which are oxidized by glucose oxidase producing hydrogen peroxide as a byproduct results in greater decolorization of MSW [108].

6. Conclusion

In the few last decades, interest has been developed in the field of bioremediation by using microbes. Several microorganisms such as bacteria and fungi, show a good ability to decolorize the effluent of the melanoidin based distillery industries. Thus, a better understanding of the microbial activities responsible for the degradation of melanoidins would contribute to enhancing the efficiency of the overall treatment system. In degradation process it would also be necessary to know the end product of melanoidins. Genetic improvement of isolates can be explored in future for improving their decolorization efficiency. Thus, it can be suggested that microbial decolorization holds promise and can be exploited to develop a cost effective, eco-friendly biotechnology package for the treatment of distillery effluent. More technically advanced research efforts are required for searching, exploiting new bacterial species and improvement of practical application to propagate the use of bacteria for bioremediation of industrial effluents. Enzymatic studies would be employed to understand the mechanism of degradation of melanoidins in future prospectives. Such organisms could be used in bioreactors for treatment of wastewaters or scaling up for enzyme productions. Broader validation of these new technologies and integration of different methods in the current treatment schemes will most likely in the near future, render these both efficient and economically viable. Finally, microbial bioremediation involves a combination of microbiologists, biotechnologists, chemists and engineers and is ideal to plug the ever widening gap between the different disciplines.

Author details

Anita Rani Santal[1] and Nater Pal Singh[2]

1 Department of Microbiology, M. D. University, Rohtak, Haryana, India

2 Centre for Biotechnology, M. D. University, Rohtak, Haryana, India

References

[1] Shweta, K. Current Trends in Bioremediation and Biodegradation. Journal of Bioremediation and Biodegradation (2012). e114.

[2] Asamudo, N. U, Daba, A. S, & Ezeronye, O. U. Bioremediation of Textile Effluent Using Phanerochaete chrysosporium. African Journal of Food, Agriculture, Nutrition and Development (2011).

[3] Tyagi, M. da Fonseca MMR, de Carvalho CC. Bioaugmentation and Biostimulation Strategies to Improve the Effectiveness of Bioremediation Processes. Biodegradation (2011). , 22(2), 231-241.

[4] Madhavi, A, & Rao, A. P. Effect of Industrial Effluent on Properties of Groundwater. Journal of Environmental Biology (2003). , 24(2), 187-192.

[5] Saravanamoorthy, M. D. Kumari BDR. Effect of Textile Wastewater on Morphophysiology and Yield on two Varieties of Peanut (Arachis hypogaea L.). Journal of Agricultural Technology (2007). , 3(2), 335-343.

[6] Naik, N. M, Jagdeesh, K. S, & Alagawadi, A. R. Microbial Decolorization of Spentwash: A Review. Indian Journal of Microbiology (2008). , 48-41.

[7] Joshi, H. C. Bio-Energy Potential of Distillery Effluents. BioEnergy News (1999). , 3-10.

[8] Pant, D, & Adholeya, A. Identification Ligninolytic Enzyme Activity and Decolorization Potential of two Fungi Isolated from a Distillery Effluent Contaminated Site. Water, Air and Soil Pollution (2007b). , 183-165.

[9] Chandra, R. Development of Microorganism for Removal of Color from Anaerobically Treated Distillery Effluent. Final technical report submitted to Department of Biotechnology New Delhi; (2003).

[10] Chandra, R, Kumar, K, & Singh, J. Impact of Anaerobically Treated and Untreated (Raw) Distillery Effluent Irrigation on Soil Microflora Growth Total Chlorophyll and Protein Contents of Phaseolus aureus L. Journal of Environmental Biology (2004). , 25(4), 381-385.

[11] Pazouki, M, Shayegan, J, & Afshari, A. Screening of Microorganisms for Decolorization of Treated Distillery Wastewater. Iranian Journal of Science and Technology (2008). , 32-53.

[12] FitzGibbon FSingh D, McMullan G, Marchant R. The Effect of Phenolic Acids and Molasses Spent Wash Concentration on Distillery Wastewater Remediation by Fungi. Process Biochemistry (1998). , 33(8), 799-803.

[13] Pant, D, & Adholeya, A. Biological Approaches for Treatment of Distillery Wastewater: A Review. Bioresource Technology (2007a). , 98-2321.

[14] Chandra, R, Bharagava, R. N, & Rai, V. Melanoidins as Major Colorant in Sugarcane Molasses Based Distillery Effluent and its Degradation. Bioresource Technology (2008). , 99-4648.

[15] Mohana, S, Desai, C, & Madamwar, D. Biodegradation and Decolorization of Anaerobically Treated Distillery Spent Wash by a Novel Bacterial Consortium. Bioresource Technology (2007). , 98(2), 333-339.

[16] Kumar, P, & Chandra, P. Decolorization and Detoxification of Synthetic Molasses Melanoidins by Individual and Mixed Cultures of Bacillus spp. Bioresource Technology (2006). , 97-2096.

[17] Wedzicha, B. L, & Kaputo, M. T. Melanoidins from Glucose and Glycine: Composition, Characteristics and Reactivity towards Sulphite Ion. Food Chemistry (1992). , 43-359.

[18] Reynolds, T. M. Chemistry of Non-Enzymatic Browning of the Reaction Between Aldoses and Amines. Advances in Food Research (1968). , 12-1.

[19] Kumar, V, & Wati, L. FitzGibbon FJ, Nigam P, Banat IM, Singh D, Marchant R. Bioremediation and Decolorization of Anaerobically Digested Distillery Spent Wash. Biotechnology Letters (1997a). , 19(4), 311-313.

[20] Kumar, V, Wati, L, Nigam, P, & Banat, I. M. MacMullan G, Singh D, Marchant R. Microbial Decolorization and Bioremediation of Anaerobically Digested Molasses Spent Wash Effluent by Aerobic Bacterial Culture. Microbios (1997b). , 89-81.

[21] Kim, S. B, Hayase, F, & Kato, H. Decolorization and Degradation Products of Melanoidins on Ozonolysis. Agricultural and Biological Chemistry (1985). , 49(3), 785-792.

[22] Ohmomo, S, Kaneko, Y, Sirianuntapiboon, S, Somachi, P, Atthasumpunna, P, & Nakamura, I. Decolorization of Molasses Wastewater by a Thermophilic Strain Aspergillus fumigatus G- Agricultural and Biological Chemistry (1987). , 2-6.

[23] Satyawali, Y, & Balakrishnan, M. Removal of Color from Biomethanated Distillery Spentwash by Treatment with Activated Carbons. Bioresource Technology (2007). , 98-2629.

[24] Ivarson, K. C, & Benzing-purdie, L. M. Degradation of Melanoidins by Soil Microorganisms Under Laboratory Conditions. Canadian Journal of Soil Science (1987). , 67-409.

[25] Kitts, D. D, Wu, C. H, Stich, H. F, & Powrie, W. D. Effect of Glucose-Lysine Maillard Reaction Products on Bacterial and Mammalian Cell Mutagenesis. Journal of Agriculture and Food Chemistry (1993). , 41(12), 2353-2358.

[26] Miyata, N, Iwahori, K, & Fujita, M. Manganese-Independent and-Dependent Decolorization of Melanoidin by Extracellular Hydrogen Peroxide and Peroxidases from

Coriolus hirsutus pellets. Journal of Fermentation and Bioengineering (1998). , 85(5), 550-553.

[27] Silvan, J. M. Lagemaat JVD, Olano A, Castillo MDD. Analysis and Biological Properties of Amino Acid Derivates Formed by Maillard Reaction in Foods. Journal of Pharmaceutical and Biomedical Analysis (2006). , 41-1543.

[28] Morales, F, & Jimnez-perez, S. Free Radical Scavenging Capacity of Maillard Reaction Products as Related to Color and Fluorescence. Food Chemistry (2001). , 72-119.

[29] Cammerer, B, Jaluschkov, V, & Kroh, L. W. Carbohydrates structures as part of the melanoidins skeleton. International Congress Series (2002). , 1245-269.

[30] Manisankar, P, Rani, C, & Vishwanathan, S. Effect of Halides in the Electrochemical Treatment of Distillery Effluent. Chemosphere (2004). , 57-961.

[31] Sirianuntapiboon, S, Zohsalam, P, & Ohmomo, S. Decolorization of Molasses Wastewater by Citeromyces spWR- Process Biochemistry (2004a). , 43-6.

[32] Migo, V. P. Del Rosario EJ, Matsumura M. Flocculation of Melanoidins Induced by Inorganic Ions. Journal of Fermentation and Bioengineering (1997). , 83(3), 287-291.

[33] Miura, M, & Gomyo, T. Formation of Blue Pigments in the Earlier Stage of Browning in the System Composed of D-Xylose and Glycine. Nippon Nogeikagaku Kaishi (1982). , 56-417.

[34] Hayase, F, Takahashi, Y, Tominaga, S, Miura, M, Gomyo, T, & Kato, H. Identification of Blue Pigment Formed in a D-Xylose-Glycine Reaction System. Bioscience, Biotechnology, and Biochemistry (1999). , 63(8), 1512-1514.

[35] Hodge, J. E. Chemistry of Browning Reactions in Model Systems. Journal of Agriculture and Food Chemistry (1953). , 1-928.

[36] Hayashi, T, & Namiki, M. Role of Sugar Fragmentation in an Early Stage Browning of Amino-Carbonyl Reaction of Sugar with Amino Acid. Agricultural and Biological Chemistry (1986). , 50(8), 1965-1970.

[37] Sowmeyan, R, & Swaminathan, G. Effluent Treatment Process in Molasses Based Distillery Industries: A Review. Journal of Hazardous Materials (2008). , 152-453.

[38] Venkat, S. M, Krishna, S. M, & Karthikeyan, J. Adsorption Mechanism of Acid-Azo Dye from Aqueous Solution on to Coal/Coal Based Sorbents and Activated Carbon: A Mechanist Study. In: Jayarama Reddy S. (ed.) Analytical Techniques in Monitoring the Environment. Tirupathi, India: Student offset printers; (2000). , 97-103.

[39] Bernardo, E. C, Egashira, R, & Kawasaki, J. Decolorization of Molasses Wastewater Using Activated Carbon Prepared from Cane Bagasses. Carbon (1997). , 35(9), 1217-1221.

[40] Satyawali, Y, & Balakrishnan, M. Wastewater Treatment in Molasses-Based Alcohol Distilleries for Cod and Color Removal: A Review. Journal of Environmental Management (2008). , 86-481.

[41] Bitton, G, Jung, K, & Koopman, B. Evaluation of a Microplate Assay Specific for Heavy Metal Toxicity. Archives of Environmental Contamination and Toxicology (1994). , 27(1), 25-28.

[42] Pandey, R. A, Malhotra, S, Tankhiwale, A, Pande, S, Pathe, P. P, & Kaul, S. N. Treatment of Biologically Treated Distillery Effluent- A Case Study. International Journal of Environmental Studies (2003). , 60(3), 263-275.

[43] Migo, V. P, Matsumura, M, Delrosario, E. J, & Kataoka, H. Decolorization of Molasses Wastewater Using an Inorganic Flocculent. Journal of Fermentation and Bioengineering (1993). , 75-438.

[44] Coca, M, Pena, M, & Gonzalez, G. Chemical Oxidation Processes for Decolorization of Brown Colored Molasses Wastewater. Ozone: Science and Engineering (2005). , 27-365.

[45] Kumaresan, T. Sheriffa Begum KMM, Sivashanmugam P, Anantharaman N, Sundaram S. Experimental Studies on Treatment of Distillery Effluent by Liquid Membrane Extraction. Chemical Engineering Journal (2003).

[46] De Wilde FGNDemineralization of a Molasses Distillery Wastewater. Desalination (1987). , 67-481.

[47] Bories, A, Sire, Y, & Colin, T. Odorous Compounds Treatment of Winery and Distillery Effluents During Natural Evaporation in Ponds. Water Science and Technology (2005). , 51(1), 129-136.

[48] Bhandari, H. C, Mitra, A. K, & Kumar, S. Crest's Integrated System: Reduction and Recycling of Effluents in Distilleries. In: Tewari P.K. (ed.) Liquid Asset Proceedings of Indo-EU Workshop on Promoting Efficient Water Use in Agro-based Industries. New Delhi, India: TERI Press; , 167-169.

[49] Gulati, N. Conservation of Resources Using Evaporation and Spray Drying Technology for Distillery and Paper Industries. In: Tewari P.K. (ed.) Liquid Asset Proceedings of Indo-EU Workshop on Promoting Efficient Water Use in Agro-Based Industries. New Delhi, India: TERI Press; (2004). , 163-166.

[50] Cortez LABPerez LEB. Experiences on Vinasse Disposalpart III: Combustion of Vinasse Fuel Oil Emulsions. Brazilian Journal of Chemical Engineering (1997). , 14(1), 9-18.

[51] Joshi, D. L. High Rate Anaerobic Treatment of Industrial Wastewater in Tropics. Thammasat International Journal of Science and Technology (1998). , 3(1), 1-7.

[52] Nandy, T, Shastry, S, & Kaul, S. N. Wastewater Management in a Cane Molasses Distillery Involving Bioresource Recovery. Journal of Environmental Management (2002). , 65-25.

[53] Bhoi ZMPMehrotra I, Shrivastava AK. Treatability Studies of Black Liquor by Upflow Anaerobic Sludge Blanket Reactor. Journal of Environmental Engineering and Science (2003). , 2(4), 307-331.

[54] Farhadian, M, Borghei, M, & Umrania, V. V. Treatment of Beet Sugar Wastewater by UAFB Bioprocess. Bioresource Technology (2007). , 98-3080.

[55] Travieso, L, Sánchez, E, Borja, R, Benítez, F, Raposo, F, Rincón, B, & Jiménez, A. M. Evaluation of a Laboratory-Scale Stabilization Pond for Tertiary Treatment of Distillery Waste Previously Treated by a Combined Anaerobic Filter-Aerobic Trickling System. EcolEngg (2006). , 27(2), 100-108.

[56] Sharma, J, & Singh, R. Effect of Nutrients Supplementation on Anaerobic Sludge Development and Activity for Treating Distillery Effluent. Bioresource Technology (2001). , 79-203.

[57] Driessen WJBMTielbaard MH, Vereijken TLFM. Experience on Anaerobic Treatment of Distillery Effluent with the UASB Process. Water Science and Technology (1994). , 30(12), 193-201.

[58] Blonskaja, V, Menert, A, & Vilu, R. Use of Two-Stage Anaerobic Treatment for Distillery Waste. Advances in Environmental Research (2003). , 7(3), 671-678.

[59] Beltran, F. J, Garcia-araya, J. F, & Alvarez, P. M. Wine Distillery Wastewater Degradation Improvement of Aerobic Biodegradation by Means of an Integrated Chemical (Ozone)-Biological Treatment. Journal of Agricultural and Food Chemistry (1999). , 47-3919.

[60] Kitamura, Y, Maekawa, T, Tagawa, A, Hayashi, H, & Farrell-poe, K. L. Treatment of Strong Organic Nitrogenous Wastewater by an Anaerobic Contact Process Incorporating Ultrafiltration. Applied Engineering in Agriculture (1996). , 12-709.

[61] Yeoh, B. G. Two-Phase Anaerobic Treatment of Cane-Molasses Alcohol Stillage. Water Science and Technology (1997). , 36-441.

[62] Harada, H, Uemura, S, Chen, A. C, & Jayadevan, J. Anaerobic Treatment of a Recalcitrant Distillery Wastewater by a Thermophilic UASB Reactor. Bioresource Technology (1996). , 55(3), 215-221.

[63] Bhatti, Z. I, Furukawa, K, & Fujita, M. Microbial Diversity in UASB Reactors. Pure and Applied Chemistry (1997). , 69(11), 2431-2438.

[64] Babu, V. L, Mohan, S. V, & Sharma, P. N. Influence of Reactor Configuration on Fermentative Hydrogen Production During Wastewater Treatment. International Journal of Hydrogen Energy (2009). , 34(8), 3305-3312.

[65] Bhatia, D, Vieth, W. R, & Vekatasubramanian, K. Steady State and Transit Behaviour in Microbial Methanification: II. Mathematical Modeling and Verification. Biotechnology and Bioengineering (1985). , 27(8), 1192-1198.

[66] Parkin, G. F, & Speece, R. E. Yang CHJ, Kocher WM. Response of Methane Formation Systems to Industrial Toxicants. Journal of the Water Pollution Control Federation (1983). , 55-44.

[67] Karhadkar, P. P, Handa, B. K, & Khanna, P. Pilot-Scale Distillery Spent Wash Biomethanation. Journal of Environmental Engineering (1990). , 116(6), 1029-1045.

[68] Krzywonos, M, & Seruga, P. Decolorization of Sugar Beet Molasses Vinasse a High Strength Distillery Wastewater by Lactic Acid Bacteria. Polish Journal of Environmental Studies (2012). , 21(4), 943-948.

[69] Krzywonos, M. Decolorization of Sugar Beet Distillery Effluent Using Mixed Cultures of Bacteria of the Genus Bacillus. African Journal of Biotechnology (2012). , 11(14), 3464-3475.

[70] Yadav, S, & Chandra, R. Biodegradation of Organic Compounds of Molasses Melanoidin (MM) from Biomethanated Distillery Spent Wash (BMDS) during the Decolorization by a Potential Bacterial Consortium. Biodegradation (2012). , 23-609.

[71] Santal, A. R, Singh, N. P, & Saharan, B. S. Biodegradation and Detoxification of Melanoidin from Distillery Effluent Using an Aerobic Bacterial Strain SAG_5 of Alcaligenes faecalis. Journal of Hazardous Materials (2011). , 193-319.

[72] Bharagava, R. N, Chandra, R, & Rai, V. Isolation and Characterization of Aerobic Bacteria Capable of the Degradation of Synthetic and Natural Melanoidins from Distillery Effluent. World Journal of Microbiology and Biotechnology (2009). , 25-737.

[73] Tondee, T, & Sirianuntapiboon, S. Decolorization of Molasses Wastewater by Lactobacillus plantarum NoPV Bioresource Technology (2008). , 71-1861.

[74] Jiranuntipon, S, Chareonpornwattana, S, Damronglerd, S, Albasi, C, & Delia, M. L. Decolorization of Synthetic Melanoidins-Containing Wastewater by a Bacterial Consortium. Journal of Industrial Microbiology and Biotechnology (2008). , 35-1313.

[75] Adikane, H. V, Dange, M. N, & Selvakumari, K. Optimization of Anaerobically Digested Distillery Molasses Spent Wash Decolorization Using Soil as Inoculum in the Absence of Additional Carbon and Nitrogen Source. Bioresourse Technology (2006). , 97-2131.

[76] Sivakumar, N, Saravanan, V, & Balakumar, S. Bacterial Decolorization and Bioremediation of an Anaerobically Digested Molasses Spent Wash. Asian Journal of Microbiology, Biotechnology and Environment Science (2006). , 8(2), 291-295.

[77] Chavan, M. N, Kulkarni, M. V, Zope, V. P, & Mahulikar, P. P. Microbial Degradation of Melanoidins in Distillery Spent Wash by an Indigenous Isolate. Indian Journal of Biotechnology (2006). , 5-416.

[78] Ghosh, M, Verma, S. C, Mengoni, A, & Tripathi, A. K. Enrichment and Identification of Bacteria Capable of Reducing Chemical Oxygen Demand of Anaerobically Treated Molasses Spent Wash. Journal of Applied Microbiology (2004). , 96-1278.

[79] Sirianuntapiboon, S, Phothilangka, P, & Ohmomo, S. Decolorization of Molasses Wastewater by a Strain of Acetogenic Bacteria. Bioresource Technology (2004b). (BP103), 92-31.

[80] Cibis, E, Kent, C. A, Krzywonos, M, Garncarek, Z, Garncarek, B, & Miskiewicz, T. Biodegradation of Potato Slops from a Rural Distillery by Thermophilic Aerobic Bacteria. Bioresource Technology (2002). , 85-57.

[81] Savant, D. V, Shouche, Y. S, Prakash, S, & Ranade, D. R. Methanobrevibacter acididurans sp. nov a Novel Methanogen from a Sour Anaerobic Digester. International Journal of Systematic and Evolutionary Microbiology (2002). , 52-1081.

[82] Ghosh, M, Ganguli, A, & Tripathi, A. K. Treatment of Anaerobically Digested Distillery Spentwash in a Two-Stage Bioreactor Using Pseudomonas putida and Aeromonas sp. Process Biochemistry (2002). , 37(8), 857-862.

[83] Dahiya, J, Singh, D, & Nigam, P. Decolorization of Molasses Wastewater by Cells of Pseudomonas fluorescens Immobilized on Porous Cellulose Carrier. Bioresource Technology (2001b). , 78-111.

[84] Jain, N, Minocha, A. K, & Verma, C. L. Degradation of Predigested Distillery Effluent by Isolated Bacterial Strains. Indian Journal of Experimental Botany (2002). , 40-101.

[85] Masaharu, S, Hiromi, K, Akihiro, N, Naoshi, F, & Rikiya, T. Screening and Identification of Lactic Acid Bacteria with the Ability to Decolorize Wastewater from a Molasses Alcohol Distillery. Journal of the Brewing Society of Japan (1998). , 93(12), 982-989.

[86] Kambe, T. N, Shimomura, M, Nomura, N, Chanpornpong, T, & Nakahara, T. Decolorization of Molasses Wastewater by Bacillus Spunder Thermophilic and Anaerobic Conditions. Journal of Bioscience and Bioengineering (1999). , 87(1), 119-121.

[87] Sirianuntapiboon, S. Selection of Acid Forming Bacteria Having Decolorization Activity for Removal of Color Substances from Molasses Wastewater. Thammasat International Journal of Science and Technology (1999). , 4(2), 1-12.

[88] Ohmomo, S, Daengsabha, W, Yoshikawa, H, Yui, M, Nozaki, K, Nakajima, T, & Nakamura, I. Screening of Anaerobic Bacteria with the Ability to Decolorize Molasses Melanoidin. Agricultural and Biological Chemistry (1988a). , 57-2429.

[89] Kumar, V, Wati, L, Nigam, P, Banat, I. M, Yadav, B. S, Singh, D, & Marchant, R. Decolorization and Biodegradation of Anaerobic Digested Sugarcane Molasses Spent-

wash Effluent from Biomethanation Plants by White Rot Fungi. Process Biochemistry (1998). , 33(1), 83-88.

[90] Miranda, P. M, Benito, G. G, Cristobal, N. S, & Nieto, C. H. Color Elimination from Molasses Wastewater by Aspergillus niger. Bioresource Technology (1996). , 57-229.

[91] Angayarkanni, J, Palaniswamy, M, & Sawminathan, K. Biotreatment of Distillery Effluent Using Aspergillus niveus. Bulletin of Environmental Contamination and Toxicology (2003). , 70-268.

[92] Shayegan, J, Pazouki, M, & Afshari, A. Continuous Decolorization of Anaerobically Digested Distillery Wastewater. Process Biochemistry (2004). , 40-1323.

[93] Guimaraes, C. Bento LSM, Mota M. Biodegradation of Colorants in Refinery Effluents-Potential Use of the Fungus Phanerochaete chrysosporium. International Sugar Journal (1999). , 101-246.

[94] Ohmomo, S, Itoh, N, Watanabe, Y, Kaneko, Y, Tozawa, Y, & Ueda, K. Continuous Decolorization of Molasses Wastewater with Mycelia of Coriolus versicolor Ps4a. Agricultural and Biological Chemistry (1985b). , 49-2551.

[95] Ohmomo, S, Kainuma, M, Kamimura, K, Sirianuntapiboon, S, Oshima, I, & Atthasumpunna, P. Adsorption of Melanoidin to the Mycelia of Aspergillus oryzae Y- Agricultural and Biological Chemistry (1988b). , 2-32.

[96] Borja, R, Martin, A. M, & Duran, M. M. Enhancement of the Anaerobic Digestion of Wine Distillery Wastewater by the Removal of Phenolics Inhibitors. Bioresource Technology (1993). , 45-99.

[97] Sirianuntapiboon, S, Sihanonth, P, Somachai, P, Atthasampunna, P, & Hayashida, S. An Adsorption Mechanism for Melanoidins Decolorization by Rhizoctonia sp. Bioscience, Biotechnology, and Biochemistry (1995). , 59-1185.

[98] Murata, M, Terasawa, N, & Homma, S. Screening of Microorganisms to Decolorize a Model Melanoidin and the Chemical Properties of a Microbially Treated Melanoidin. Bioscience, Biotechnology, and Biochemistry (1992). , 56(8), 1182-1187.

[99] Kida, K, Morimura, S, Abe, N, & Sonoda, Y. Biological Treatment of Shochu Distillery Wastewater. Process Biochemistry (1995). , 30(2), 125-132.

[100] Fujita, M, Era, A, Ike, M, Soda, S, Miyata, N, & Hirao, T. Decolorization of Heat-Treatment Liquor of Waste Sludge by a Bioreactor Using Polyurethane Foam-Immobilized White Rot Fungus Equipped with an Ultramembrane Filtration Unit. Journal of Bioscience Bioengineering (2000). , 90(4), 387-94.

[101] Fumi, M. D, Parodi, G, Parodi, E, & Silva, A. Optimisation of Long Term Activated Sludge Treatment of Winery Wastewater. Bioresource Technology (1995). , 52(1), 45-51.

[102] Benito, G. G, Miranda, M. P, & Santos, D. R. Decolorization of Wastewater from an Alcoholic Fermentation Process with Trametes Versicolor. Bioresource Technology (1997)., 61-33.

[103] Sirianuntapiboon, S, & Chairattanawan, K. Some Properties of Coriolus sp. No.20 for Removal of Color Substances from Molasses Wastewater. Thammasat International Journal of Science and Technology (1998)., 3(2), 74-79.

[104] Miyata, N, Mori, T, Iwahori, K, & Fujita, M. Microbial Decolorization of Melanoidin-Containing Wastewaters: Combined Use of Activated Sludge and the Fungus Coriolus hirsutus. Journal of Bioscience and Bioengineering (2000)., 89(2), 145-150.

[105] Dahiya, J, Singh, D, & Nigam, P. Decolorization of Synthetic and Spentwash Melanoidins Using the White-Rot Fungus Phanerochaete chrysosporium JAG-40. Bioresource Technology (2001a)., 78-95.

[106] Malandra, L, Wolfaardt, G, Zietsman, A, & Viljoen-bloom, M. Microbiology of a Biological Contractor for Winery Wastewater Treatment. Water Research (2003)., 37-4125.

[107] Kwak, E. J, Lee, Y. S, Murata, M, & Homma, S. Effect of Reaction pH on the Photodegradation of Model Melanoidins. Lebensmittel-Wissenschaft und-Technologie (2004)., 37-255.

[108] Raghukumar, C, Mohandass, C, Kamat, S, & Shailaja, M. S. Simultaneous Detoxification and Decolorization of Molasses Spent Wash by the Immobilized White-Rot Fungus Flavodon flavus Isolated from a Marine Habitat. Enzyme and Microbial Technology (2004)., 35-197.

[109] Raghukumar, C, & Rivonkar, G. Decolorization of Molasses Spent Wash by the White-Rot Fungus Flavodon flavus Isolated from a Marine Habitat. Applied Microbiology and Biotechnology (2001)., 55-510.

[110] Potentini, M. F, & Rodríguez-malaver, A. J. Vinasse Biodegradation by Phanerochaete chrysosporium. Journal of Environmental Biology (2006)., 27(4), 661-665.

[111] Thakkar, A. P, Dhamankar, V. S, & Kapadnis, B. P. Biocatalytic Decolorization of Molasses by Phanerochaete chrysosporium. Bioresource Technology (2006)., 97-1377.

[112] Sirianuntapiboon, S, Somachi, P, & Sihanonth, P. Ohmomo PAS. Microbial Decolorization of Molasses Wastewater by Mycelia Sterilia D90. Agricultural and Biological Chemistry (1988)., 52(2), 393-398.

[113] Seyis, I, & Subasioglu, T. Screening of Different Fungi for Decolorization of Molasses. Brazilian Journal of Microbiology (2009)., 40-61.

[114] Kaushik, G, & Thakur, I. S. Isolation of Fungi and Optimization of Process Parameters for Decolorization of Distillery Mill Effluent. World Journal of Microbiology and Biotechnology (2009)., 25-955.

[115] Valderrama, L. T. Del Campo CM, Rodriguez CM, Bashan LE, Bashan Y. Treatment of Recalcitrant Wastewater from Ethanol and Citric Acid Using the Microalga Chlorella vulgaris and the Macrophyte Lemna minuscule. Water Research (2002). , 36-4185.

[116] Kalavathi, D. F, Uma, L, & Subramanian, G. Degradation and Metabolization of the Pigment-Melanoidin in Distillery Effluent by the Marine Cyanobacterium Oscillatoria boryana BDU 92181. Enzyme and Microbial Technology (2001). , 29-246.

[117] Patel, A, Pawar, P, Mishra, S, & Tewari, A. Exploitation of Marine Cyanobacteria for Removal of Color from Distillery Effluent. Indian Journal of Environmental Protection (2001). , 21-1118.

[118] Moriya, K, Lefuji, H, Shimoi, H, Sato, S, & Tadenuma, M. Treatment of Distillery Wastewater Discharged from Beet Molasses-Spirits Production Using Yeast. Journal of Fermentation and Bioengineering (1990). , 69-138.

[119] Shojaosadati, S. A, Khalilzadeh, R, Jalilzadeh, A, & Sanaei, H. R. Bioconversion of Molasses Stillage to Protein as an Economic Treatment of this Effluent. Resources, Conservation and Recycling (1999).

[120] Karam, J, & Nicell, J. A. Potenital Applications of Enzymes in Waste Treatment. Journal of Chemical Technology and Biotechnology (1997). , 69-141.

[121] Sangave, P. C, & Pandit, A. B. Enhancement in Biodegradability of Distillery Wastewater Using Enzymatic Pretreatment. Journal of Environmental Management (2006a). , 78-77.

[122] Klibanov, A. M, Alberti, B. N, Morris, E. D, & Felshin, L. M. Enzymatic Removal of Toxic Phenols and Anilines from Wastewaters. Journal of Applied Biochemistry (1980). , 2-414.

[123] Klibanov, A. M, & Morris, E. D. Horseradish Peroxidases for the Removal of Carcinogenic Aromatic Amines from Water. Enzyme and Microbial Technology (1981). , 3-119.

[124] Aitken, M. D, & Irvine, R. L. Stability Testing of Ligninase and Mnperoxidase from Phanerochaete Chrysosporium. Biotechnology and Bioengineering (1989). , 34(10), 1251-1260.

[125] Duff, S. J, Moritz, J. W, & Andersen, K. L. Simultaneous Hydrolysis and Fermentation of Pulp Mill Primary Clarifier Sludge. Canadian Journal of Chemical Engineering (1994). , 72-1013.

[126] Ferrer, I, Dezotti, M, & Duran, N. Decolorization of Kraft Effluent by Free and Immobilized Lignin Peroxidases and Horse Radish Peroxidases. Biotechnology Letters (1991). , 13-577.

[127] Dec, J, & Bollag, J. M. Use of Plant Material for the Decontamination of Water Polluted with Phenols. Biotechnology and Bioengineering (1994). , 44-1132.

[128] Watanabe, Y, Sugi, R, & Tanaka, Y. Enzymatic Decolorization of Melanoidin by Coriolus sp. No 20. Agricultural and Biological Chemistry (1982). , 46(6), 1623-1630.

[129] Ohmomo, S, Aoshima, I, Tozawa, Y, Sakurada, N, & Ueda, K. Purification and Some Properties of Melanoidin Decolorizing Enzymes and P-4 from Mycelia of Coriolus versicolor Ps4a. Agricultural and Biological Chemistry (1985a). , 3.

[130] Arora, D. S, Chander, M, & Gill, P. K. Involvement of Lignin Peroxidase, Manganese Peroxidase and Laccase in Degradation and Selective Ligninolysis of Wheat Straw. International Biodeterioration and Biodegradation (2002). , 50-115.

[131] Rubia TDLLinares A, Perez J, Dorado JM, Romera J, Martinez J. Characterization of Manganese-Dependent Peroxidase Isoenzymes from the Ligninilytic Fungus Phanerochaete flavido-alba. Research in Microbiology (2002). , 153-547.

[132] Mansur, M, Suarez, T, & Fernandez-larrea, J. B. Brizuela MAA, Gonzalezi AE. Identification of a Laccase Gene Family in the New Lignin-Degrading Basidiomycetes CECT 20197. Applied and Environmental Microbiology (1997). , 63(7), 2637-2646.

[133] Dehorter, B, & Blondeau, R. Isolation of an Extracellular Mn-Dependent Enzyme Mineralizing Melanoidins from the White Rot Fungus Trametes versicolor. FEMS Microbiology Letters (1993). , 109(1), 117-122.

[134] Lee, T. H, Aoki, H, Sugano, Y, & Shoda, M. Effect of Molasses on the Production and Activity of Dye-Decolorizing Peroxidase from Geotrichum candidum. Journal of Bioscience and Bioengineering (2000). , 89-545.

[135] Souza, D, Tiwari, D. T, Sah, R, & Raghukumar, A. K. C. Enhanced Production of Laccase by a Marine Fungus During Treatment of Colored Effluents and Synthetic Dyes. Enzyme and Microbial Technology (2006). , 38-504.

[136] Sangave, P. C, & Pandit, A. B. Ultrasound and Enzyme Assisted Biodegradation of Distillery Wastewater. Journal of Environmental Management (2006b). , 80-36.

[137] Yadav, S, Chandra, R, & Rai, V. Characterization of Potential MnP Producing Bacteria and its Metabolic Products During Decolorization of Synthetic Melanoidins due to Biostimulatory Effect of D-Xylose at Stationary Phase. Process Biochemistry (2011). , 86-1774.

[138] Raghukumar, C, Souza, D, Thorn, T. M, Reddy, R. G, & Lignin-modifying, C. A. Enzymes of Flavodon flavus a Basidiomycete isolated from a Coastal Marine Environment. Applied Environmental Microbiology (1999). , 65-2103.

[139] Raghukumar, C. Fungi from Marine Habitat: Application in Bioremediation. Mycological Research (2000). , 104-1222.

Chapter 6

Degradability: Enzymatic and in Simulated Compost Soil of PLLA:PCL Blend and on Their Composite with Coconut Fiber

Yasko Kodama

Additional information is available at the end of the chapter

http://dx.doi.org/10.5772/56231

1. Introduction

The problem of non-biodegradable plastic waste remains a challenge due to its negative environmental impact. In this sense, poly(L-lactic acid), PLLA, and poly(ε-caprolactone), PCL, have been receiving much attention lately due to their biodegradability in human body as well as in the soil, biocompatibility, environmentally friendly characteristics and non-toxicity [1-5]. PLLA is a poly(α-hydroxy acid) and PCL is a poly(ω-hydroxy acid) [1]. PLLA is a hard, transparent and crystalline polymer. On the other hand, PCL can be used as a polymeric plasticizer because of its ability to lower elastic modulus and to soften other polymers [6]. The original reasons for preparing polymer blends are to reduce costs by combining high-quality polymers with cheaper materials (although this approach is usually accompanied by a drastic worsening of the properties of the polymer) and to create a polymer that has a desired combination of the different properties of its components. However, according to Michler [7] usually different polymers are incompatible. Improved properties can be only realized if the blend exhibits optimum morphology. According to Sawyer et al. [8], in polymer science, the term morphology generally refers to form and organization on a size scale above the atomic arrangement, but smaller than the size and shape of the whole sample. Thus, improving compatibility between the different polymers and optimizing the morphology are the main issues to address when producing polymer blends [3]. Moreover, both polymers PLLA and PCL can be used in biomedical applications, which require a proper sterilization process. Nowadays, the most suitable sterilization method is high energy irradiation. However, it is important to remind that polymeric structural changes are induced by radiation processing of polymers, such as scission and crosslinking [9-12]. According to the principles of radiation chemistry, very reactive intermediate, free radicals, ions and excited states are formed when macromolecules of polymers are submitted to ionizing radiation, where they are then free to react with one another or initiate further reactions among the polymeric chains, thus giving rise to changes in material properties. These intermediates can follow several reactions paths that result in disproportion, hydrogen abstraction, arrangements and/or formation of new bonds. The combination of two radicals leads to cross-linking or recombination in the amorphous and crystalline regions, respectively, whereas chain transfer and the subsequent splitting results in chain scission. Usually both these processes take place simultaneously for many polymers [10,11].

The morphology of the blends affects the thermo mechanical properties as well as the biodegradation of the polymers. In particular, surface structure and morphology of the biodegradable polymer blends have a great impact on the enzymatic degradation behavior. The development of polymeric materials susceptible to microbiological degradation and that have similar performance to conventional polymers has been intensely studied. The intention would be that those materials reduce waste volume while suffer degradation in sanitary waste deposit, or they could be treated in composting plants [13]. Enzymatic and non-enzymatic degradations occur easier in the amorphous region [14,15]. Kikkawa et al. [16] cited that one of the approaches used to generate biodegradable materials with a wide range of physical properties is blending, and miscibility of blends is one of the most important factors affecting the final polymer properties.

Nishino et al. [17] cited that cellulose is the most abundant form of biomass and the form most likely to be used as reinforcement fibers, not only because of ecological and economic reasons, but also because of their high mechanical and thermal performance. Thus, incorporating fibers of low cost to the polymeric blend, it is possible to obtain an improvement of the mechanical properties without loss of the original characteristics of polymeric components. Regarding the irradiation effects, vegetable fiber, like as coconut fiber, is composed by cellulose and lignin, which suffer chemical alteration by irradiation such as scission or cross-linking. In the case of natural polymers, such as cellulose, main chain scission occurs predominantly due to irradiation and as a result molecular weight' decrease [10].

Liu et al. and Lenglet et al. [18,19] cited that biodegradability of PCL and PLLA has also been investigated under environmental conditions. The controlled degradation of polymers is sometimes desired for biomedical applications, besides the environmental purposes [7]. It has been seen that PLLA is bio absorbable, that is, the hydrolytic degradation by-products formed can be fully assimilated by microorganisms such as fungi or bacteria. On the other hand, PCL is promptly biodegraded by environmental microorganisms. So, both PCL and PLLA can be considered as environmentally friendly polymers.

Kolybaba et al. [20] mentioned that biodegradable plastics are those that undergo significant enough modification on their chemical structure under specific environmental condition. Those changes result on mechanical and physical properties losses that are measurable by standard methods of testing. Biodegradable plastics suffer degradation under action of microorganisms that has natural occurrence, for instance, bacteria, fungi and algae. The plastic engineered to be entirely biodegradable is classified within the main classes of polymeric materials. In this category, polymeric matrix can be from natural resources and reinforcement

http://dx.doi.org/10.5772/56231

fibers would be obtained from vegetal fibers. So, microorganisms are able to consume completely those materials, eventually releasing carbon dioxide and water as by-products [20]. PCL, PLLA and coconut fiber composites studied in this chapter may be categorized in that class.

According to Müller [13], there are different approaches concerning the type of test to be applied to evaluate degradation of polymeric materials in the environment and, also, what conclusion can be obtained from that. As principle, tests can be divided in three categories, field test, simulation and laboratorial tests. Nevertheless field test, for instance, in which samples are buried on the ground, or putting them in a lake or river, or performing general process of composting of polymeric biodegradable material, represent the ideal practical conditions. There are several disadvantages associated to this kind of test. One of the problems would be to control environment conditions like temperature, pH, or humidity. Another point to be considered is to analytically monitor the degradation process, in most cases it would be possible to visually evaluate alterations of the sample, or maybe evaluate the disintegration by measuring weight loss. Most reproducible tests are laboratorial ones, well-defined medium and, inoculated with specific microorganisms to a particular polymer are utilized. In those cases, enzymatic activity is optimized to a particular microorganism and, frequently present more elevated degradation rate than the ones observed in natural conditions. This is considered as an advantage to the study of basic mechanism of polymer biodegradation. Although results lead to limited conclusion related to real degradation rate on the natural environment, those tests have widely been used.

2. Material and method

2.1. Material

Samples of PCL and PLLA homopolymers; PCL:PLLA 20:80 (w:w) blend; and composites of the blend containing 5% and 10% of coconut fiber (chemically untreated and acetylated) were prepared in triplicate.

2.2. Coconut fiber

Coir coconut fibers for composite preparation were kindly provided by Embrapa – Paraipaba region, Ceará.

Size reduction of the coconut fibers was carried out using helix mill Marconi – modelo MA 680, from Laboratório de Matéria-prima Particulados e Sólidos Não Metálicos – LMPSol, Departamento de Engenharia de Materiais of Escola Politécnica/USP.

The fiber size distribution was measured using sieves of the Tyler series 16, 20, 35 and 48, fiber sizes of 1.0mm, 0.84mm, 0.417mm, and 0.297mm, respectively. The 0.297-0.417mm fibers size was used for the assays. The triturated material was separated using a sieve shaker Produtest, for 1 min.

In order to remove lignin from coconut fiber surface, fibers were soaked with Na_2SO_3 2% aqueous solution for 2h using ultrasound. Coconut fibers were washed several times with tap water and finally, tree times with deionized water, as described in the literature [21].

Coconut fiber acetylation was performed as described by d'Almeida et al. [22]. As received fibers from Embrapa were soaked in a solution of acetic anhydrate and acetic acid (1.5:1.0, w:w). It was used as a catalyst, 20 drops of sulfuric acid in 500mL solution. Those groups of sets were submitted to ultrasound for 3h, then for more 24h rest at the same solution. Fibers were washed using tap water and for more 24h rested in deionized water. Fibers were separated from water and washed with acetone, after that, were evaporated at room temperature.

2.3. Preparation of composite pellets and sheets

PCL (pellets, $\overline{M}_w$ =2.14•10^5 g•mol^{-1}; $\overline{M}_w$ / $\overline{M}_n$= 1.423), PLLA (pellets, $\overline{M}_w$=2.64•10^5 g•mol^{-1} $\overline{M}_w$ / $\overline{M}_n$= 1.518 – Gel Permeation Chromatographic values) and dried coconut fiber (from Embrapa – Empresa Brasileira de Pesquisa Agropecuária, Ceará, Brazil) were used to prepare blends and composites. A Labo Plastomil model 50C 150 of Toyoseiki twin screw extruder was used for pellets preparation. Pellets of PCL:PLLA 20:80 (w:w) blend and composites containing 5 and 10% of untreated and chemically treated coconut fiber were prepared at AIST.

Sheets (150mm x 150mm x 0.5mm) of PCL, PLLA, PCL:PLLA 20:80 (w:w) blend and composites containing 5 and 10% untreated and chemically treated coconut fiber were prepared using Ikeda hot press equipment of Japan Atomic Energy Agency, JAEA. Mixed pellets of samples were preheated at 195°C for 3 min and then pressed by under heating at the same temperature for another 3 min under pressure of 150 kgf• cm^{-2}. Samples sheets were then cooled in the cold press using water as a coolant for 3 min.

For degradability tests, samples were taken from hot compressed polymeric sheets, cut into 15mm × 15mm pieces. Non-irradiated and, electron beam (EB), irradiated samples with absorbed doses of 50 kGy and 100 kGy were studied.

2.4. Electron beam irradiation

Irradiation was performed at JAEA using electron beam accelerator (2 MeV; 2 mA), absorbed doses of 50 and 100 kGy, dose rate of 0.6 kGy s^{-1}. The energy and current parameters condition of irradiation were enough to the electron beam goes through the 0.5mm thickness sheets.

Absorbed dose is the amount of energy absorbed per unit mass of irradiated material. The SI unit for absorbed dose is joules per kilogram (J kg^{-1}), which is given the special name gray (symbol, Gy). The absorbed dose rate is the absorbed dose per unit time and has the units gray per unit time, for instance kGy s^{-1}. The absorbed dose is a direct measure of the energy transferred to the irradiated material that is capable of producing chemical or physical change [23].

http://dx.doi.org/10.5772/56231

3. Method

3.1. Enzymatic degradation

A buffer solution with phosphate, pH 7, and lipase enzyme obtained from *Pseudomonas cepacia*, of Aldrich, was prepared. Solution concentrations were kept at 35 unities of enzymatic activity. Flasks were maintained in hot water bath at 37°C. System buffer-enzyme was preserved by 7 days (168h). Samples were exposed to enzymatic action for 0, 24, 72, 120 and 168 hours. After enzymatic exposure, samples were washed with water, dried and weighted (mass retention determination). Tests were performed in duplicate, subtracted test control without enzyme.

3.2. Biodegradability in soil

Samples of approximately 10mm × 10mm were buried in plastic trays containing simulated compost soil previously prepared, with 23 % humus, 23 % organic material (tree leaves, coffee powder, food waste and cattle manure), 23 % sand and distilled water to complete 100 %. Simulated compost was characterized for nitrogen and total carbon content, ABNT 1167 and, pH. Simulated compost soil characterization results: pH 7.8; humidity 30 ± 10 %; total carbon 18.4 %; total nitrogen 0.83 %. Samples were removed from the soil at 30, 60, 90 e 120 days of ageing. After that, they were mechanically cleaned, and dried at room temperature for 24 hours.

4. Results and discussion

4.1. Enzymatic degradation

Enzymatic degradation was performed using lipase enzyme obtained from *Pseudomonas cepacia*. In Fig. 1 it can be observed mass retention variation of non-irradiated and electron beam (EB) irradiated samples through time of degradation of homopolymers, blend and composites. Degradation rate of PCL is higher than PLLA in *Pseudomonas* lipase presence, in agreement with observed by Liu et al. [18] that lipase degrade both crystalline and amorphous PCL. According to Liu et al., enzymatic degradation of PCL has been investigated, mainly in presence of lipase enzyme. It is well known that morphology and its alteration plays an important role on hydrolytic degradability of aliphatic polyesters [18]. When the subject is enzymatic degradation, situation complicates due to specificity of enzymes. *Pseudomonas* lipase is able to break esters linkages in hydrophobic substrates, as it is PCL case. Also, it was described that PCL did not absorb water, by the other side, PLLA absorbed 2% water within 72h. The authors also informed that degradation rate of this polymer is higher in proteinase K than in lipase (8% against 1%). However, in the study of this chapter it was observed that PCL degraded approximately 30%, and PLLA 16%, at the same period of time. According to Tsuji and Ishizaka, no alteration was observed on the molar weight distribution, either mass loss of pure PLLA studied films, indicating that enzymatic hydrolysis effect caused by *Pseudomonas* lipase on the main chain of PLLA on the bulk was not significant. This confirms that enzymatic degradation occurs preferably on the surface of the sample [15].

Calil et al. and Sivalingam et al. [24,25] cited that the presence of one polymer affects degradability of the other polymer. Lenglet et al. observed that PLLA addition to PCL reduced drastically degradation of PCL of the blends in lipase presence [19]. In the study presented in this chapter, it was possible to observe that the presence of PLLA reduced enzymatic degradation of PCL of PCL:PLLA 20:80 (w:w) blend, and after 120 hours, mass retention variation moved toward of pure PLLA behavior. Tsuji and Ishizaka [15] studied enzymatic degradation of PCL:PLLA blends using *Rhizopus arrhizus* lipase. They observed that enzymes obtained from fungi cause selective hydrolysis and PCL removal of PCL:PLLA blends without significant PLLA degradation in soil. They also cited that enzymatic degradation of PCL:PLLA in presence of *Pseudomonas* lipase and Proteinase K occurred in the interface of two polymeric phases both on the bulk and on the surface of the sample.

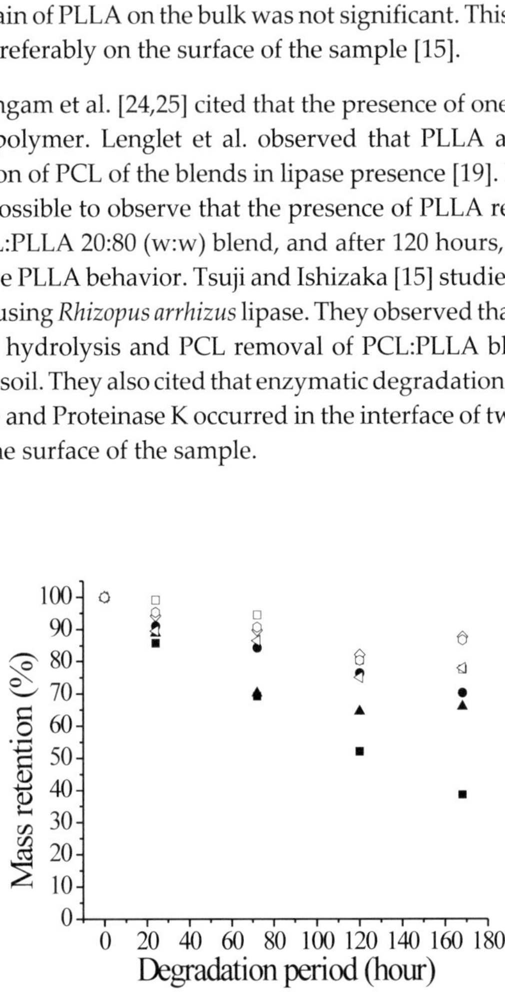

Figure 1. Mass retention variation versus degradation time, enzymatic method, of samples: (■) PCL; (●) PLLA; (▲) PCL:PLLA 20:80 (w:w); (□) composite with 5% of untreated fiber; (◇) composite with 10% of untreated fiber; (◁) composite with 5% of acetylated fiber; and (○) composite with 10% of acetylated fiber, of non-irradiated samples.

In the study of this chapter, composites degraded in a way similar of the blend through time. Mass retention values observed were higher than the blends during the same period of time of test, suggesting that coconut fibers did not significantly enzymatic degraded in this test condition. Furthermore, acetylation did not affect enzymatic degradation of composites significantly.

Tsuji and Ishizaka [15] observed that crystallinity of PCL on blends films did not change with composition variation during degradation, suggesting that this property did not affect enzymatic hydrolysis rate of PCL of the blend as it did not altered during process. Rate of enzymatic hydrolysis of blends was lower than pure PCL, suggesting that PLLA interfered on PCL hydrolysis catalyzed by lipase. One reason postulated by the authors to the deceleration of PCL degradation on the blends would be the disturbance caused by superficial adsorption of enzyme molecules on the polymeric films or by slow hydrolytic scission of main chains of PCL by molecules enzymes on the presence of PLLA molecules on the blends.

Fig. 2 shows points of lipase enzyme attack on polyesters proposed by [25].

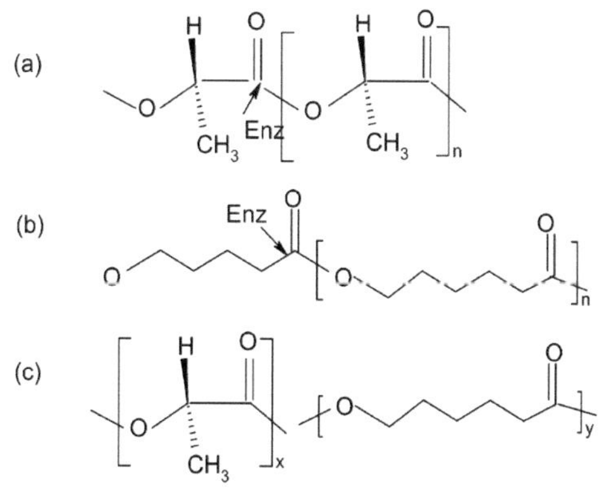

Figure 2. Points of enzyme attack on polyesters by lipase: a) PLA; b) PCL; c) PCL:PLA 14:86 (extracted from Sivalingam et al. [25])

On Fig. 3 it is possible to observe effect of radiation dose on enzymatic degradation of PCL samples irradiated with electron beam.

According to Cottam et al. [26], degradability rate of PCL irradiated with 25 kGy decreased, attributed to irradiation process. Authors cited that lipase catalyzes hydrolysis of carbonyl group linkage and one oxygen atom in the case of fat. It is the same linkage that is broken during PCL hydrolysis. They attributed that degradability rate of PCL was affected by crosslinking occurred due to irradiation. In this study, PCL irradiated with 50 kGy suffered a slight decrease on degradation rate, in agreement with authors' observation. However, PCL samples irradiated with 100 kGy presented a certain increase on degradation rate. This fact probably is related to crystallinity decrease of around 6% observed by Differential Scanning Calorimetry, DSC, of irradiated samples. On Fig. 4 it is possible to observe the effect of radiation dose on enzymatic degradation of PLLA samples irradiated with electron beam.

According to Maharana et al. [27], enzymatic degradation occurs only on the surface of a solid substrate by erosion on the surface and by weight loss, because enzymes cannot penetrate a solid polymeric substrate. Enzymes degrade selectively amorphous regions or less ordered that allows them to diffuse through substrate and, subsequently, crystalline regions are eventually degraded. In this process, molar weight and molar weight distribution of non-degraded solid substrate do not change during enzymatic degradation because only the polymer on the surface of substrate is degraded and products of low molar weight from degradation are removed of substrate by solubilization on the surrounding aqueous medium.

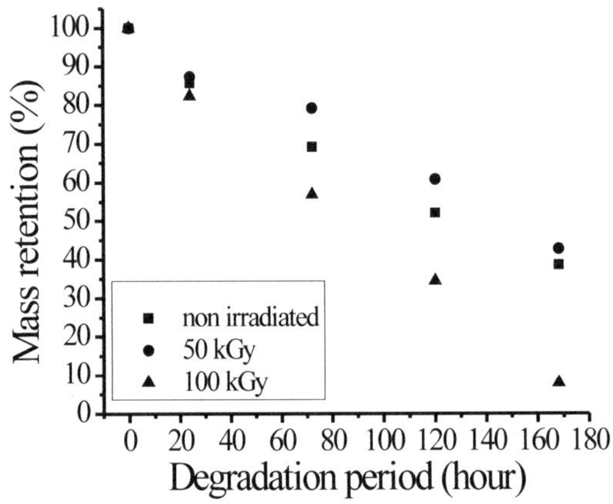

Figure 3. Mass retention variation versus degradation, enzymatic method, PCL of non-irradiated and irradiated samples with EB, radiation doses of 50 kGy and 100 kGy

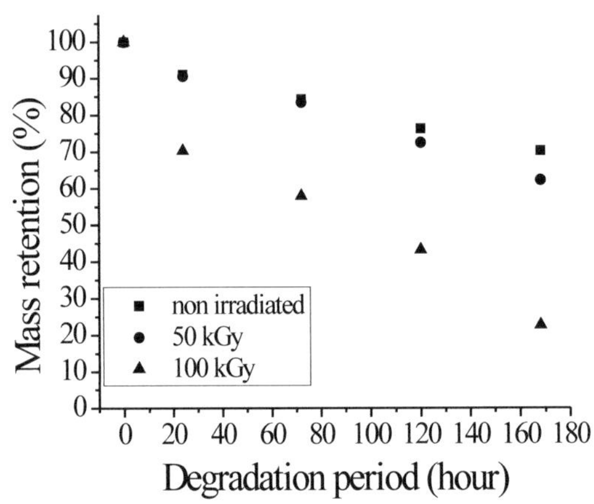

Figure 4. Mass retention variation versus degradation period, enzymatic method, PLLA samples non irradiated and irradiated with EB, radiation doses of 50 kGy and 100 kGy

There are two kinds of degradation based on the point of cleavage. Cleavage can occur in random points along polymeric chain (degradation endo-type) or at the end terminal of main chain (degradation exo type). Degradation process of lipases is based on endo type scission, so it does not depend on molar weight and on molar weight distribution. Fig. 5 shows PLA hydrolysis reaction cited by [27].

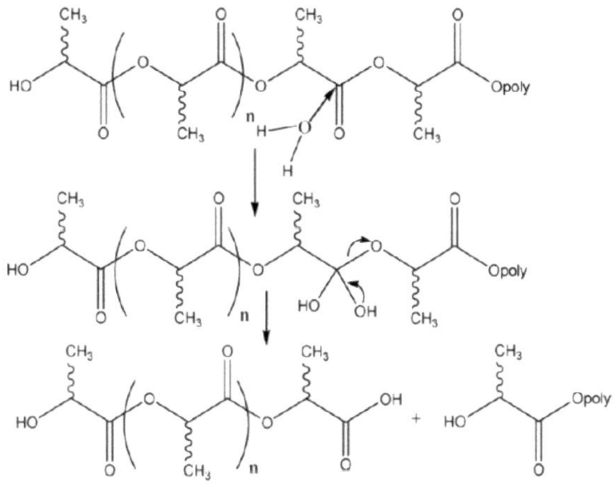

Figure 5. Hydrolysis of PLA [27].

On Fig. 6 it can be observed mass retention variation through period of enzymatic degradation of PCL:PLLA 20:80 (w:w) non-irradiated and irradiated with electron beam with absorbed radiation doses of 50 kGy and 100 kGy. Degradation values observed in this study were lower compared to the ones found in the literature, probably due to the fact that studied samples were physical mixtures of polymers. Lenglet et al. [19] studied enzymatic degradation of PCL:PLLA copolymers with $\overline{M}_n$ of 29000 to 44000, using *Pseudomonas* lipase. Authors observed that degradation occurred faster with increasing amount of PCL, attaining approximately 99% for PCL:PLA 75:25 after 72h. They suggested that PCL homopolymer can suffer degradation in presence of *Pseudomonas* lipase while PLA did not degrade in the same conditions.

In Fig. 6 irradiated blend with absorbed dose of 50 kGy presented slight reduction of degradation rate compared to non-irradiated blend, similar to the observed for homopolymers. Irradiated sample with 100 kGy showed slight increase of degradation rate after 120 hours, and then, little degradation is observed.

In Fig. 7 it is shown mass retention variation versus degradation period, by enzymatic method, of composites with 5% of chemically untreated coconut fiber, non-irradiated and EB irradiated with absorbed doses of 50 kGy and 100 kGy.

Chemically untreated fiber incorporation caused slight reduction of degradation comparing blend to composite. Probably, it is due to the fact that fibers take more time to degrade. Even though the method used by Salazar and Leão [28] was different from the one used in this study, they observed that fresh coconut fiber degraded 10% in 912 hours (38 days) by immediate degradability test, by measuring carbon dioxide release in open system, in which organic substance is subjected to metabolizing of microorganism mixture culture from environment. This carbon source of the substance can be fully consumed by microorganism metabolism into CO_2 and H_2O. It is possible to predict theoretically total CO_2 production for full biodegradation, knowing initial carbon content.

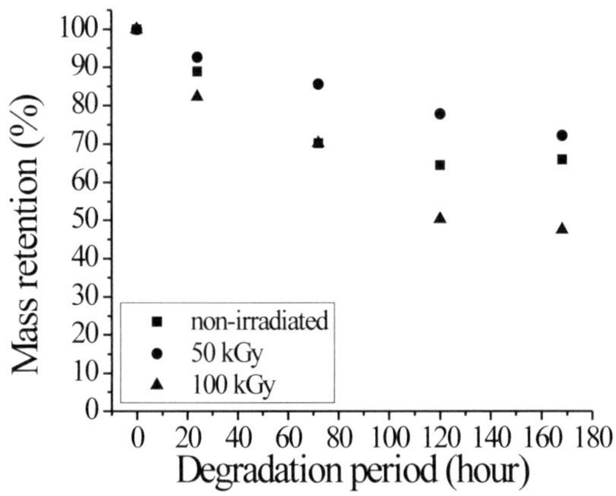

Figure 6. Mass retention variation with period of degradation, enzymatic method, of PCL:PLLA 20:80 (w:w) non-irradiated and EB irradiated samples with 50 kGy and 100 kGy absorbed doses.

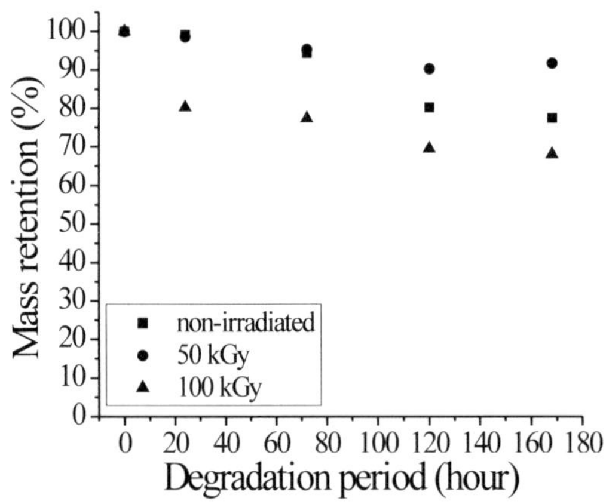

Figure 7. Mass retention variation versus degradation period, enzymatic method, composites with 5% of chemically untreated coconut fiber, non-irradiated and EB irradiated with absorbed doses of 50 kGy and 100 kGy.

In Fig. 7 it is observed that degradation rate of composites behaves in a way similar to irradiated blends, this suggests that fiber presence does not affect this parameter. Mass retention values of samples studied in 168 hour probably were affected by water absorption by PLLA and/or coconut fibers.

In Fig. 8 it is observed that when fiber content of chemically untreated coconut fiber increases in the composite, degradation rate suffers slight reduction and degradation decreases with increasing radiation dose.

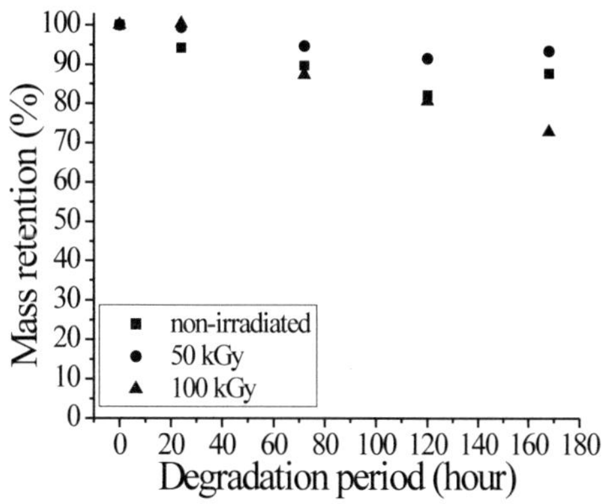

Figure 8. Mass retention variation versus degradation period, enzymatic method, composites samples with 10% of chemically untreated coconut fiber non-irradiated and EB irradiated with absorbed doses of 50 kGy and 100 kGy.

Fig. 9 shows mass retention variation with degradation period increase, enzymatic method, of composites samples with 5% acetylated fibers non-irradiated and EB irradiated with 50 kGy and 100 kGy.

Acetylation process of coconut fiber did not affect significantly degradation rate of irradiated samples. Irradiated samples suffered slight decrease of degradation rate compared to composites containing 5% of acetylated fibers, non-irradiated.

Increase of acetylated fiber content up to 10% did not affect significantly degradation neither degradation rate of EB irradiated composites with 50 kGy and 100 kGy. It was observed slight increase on mass after 168h, probably due to water absorption by PLLA or coconut fibers, Fig. 10.

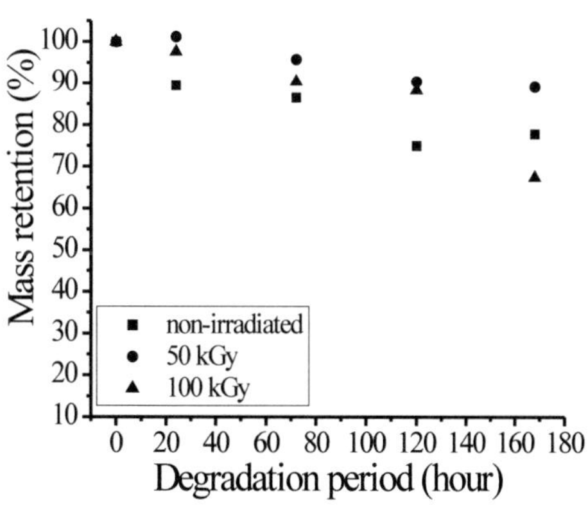

Figure 9. Mass retention variation versus degradation period, enzymatic method, and composites samples containing 5% acetylated fiber, non-irradiated and EB irradiated with absorbed doses of 50 kGy and 100 kGy.

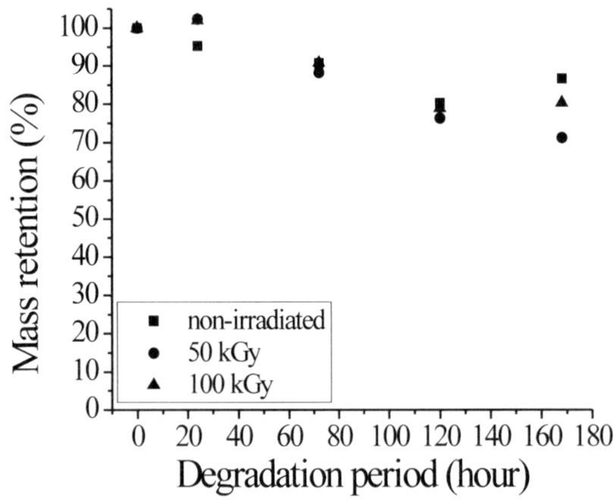

Figure 10. Mass retention variation versus degradation period, enzymatic method, of composites samples containing 10% of acetylated fiber, non-irradiated and EB irradiated with absorbed doses of 50 kGy and 100 kGy

4.2. Biodegradation in simulated compost soil

Mass retention variation versus degradation period in simulated compost soil, of non-irradiated samples PCL, PLLA, PCL:PLLA 20:80 (w:w) blend, composites wth 5% and 10% of chemically untreated fiber and composites containing 5% and 10% of acetylated fiber are shown in Fig. 11. It can be observed that all samples suffer degradation in the period of time studied. Values vary in between of 36% and 10% in 120 days for PLLA and composite with 5% of chemically untreated coconut fiber, respectively.

According to Alauzet et al. [29], PLA ester hydrolysis in abiotic aqueous media depends on autocatalysis via chain end carboxylic groups and diffusion reaction phenomena involving water and oligomer molecules formed by degradation by means of its solubility in aqueous media. When submitted to heat and water, high molecular weight PLLA degrade to oligomer (PLA of low molecular weight), dimer and monomer of lactic acid. That would explain the reason why PLA degrade in humid medium and room temperature, like organic compost or humus.

In the study of this chapter, PLLA biodegradation in simulated compost soil presented degradation rate higher than PCL, different from the behavior observed in enzymatic method, probably due to used enzyme specificity in the assay.

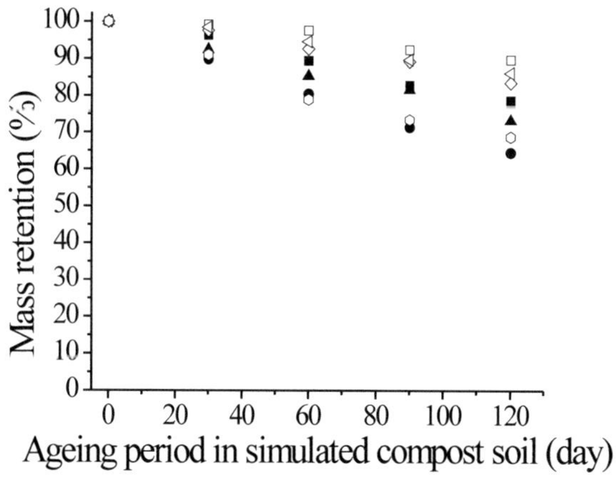

Figure 11. Mass retention variation versus biodegradation period in simulated compost soil, of samples: (■) PCL; (●) PLLA; (▴) PCL:PLLA 20:80 (w:w); (□) composite with 5% chemically untreated coconut fiber; (◇) composite with 10% chemically untreated coconut fiber; (◁) composite with 5% acetylated fiber; and (○) composite with 10% of acetylated fiber

Effect of radiation dose on the mass retention of EB irradiated samples with doses of 50 kGy and 100 kGy of PCL, is presented in Fig.12; of PLLA, in Fig.13; of blend PCL:PLLA 20:80 (w:w), in Fig.14; of composite containing 5% of chemically untreated fiber, in Fig.15; of composite with 10% of chemically untreated fiber, in Fig.16; of composite with 5% of acetylated fiber, in Fig.17; and of composite with 10% of acetylated fiber, in Fig.18.

Lotto et al. [30] observed that PCL did not suffer degradation in compost soil at room temperature even after 300 days. However, after temperature increase up to 46°C, it was observed by the authors 36% weight loss of PCL samples in 120 days. This fact was attributed to non-enzymatic hydrolysis of esters bonds due to temperature increase, that condition favored microorganism action that exists in natural soil and uses polymers as nutrient.

In this study, it was observed that at room temperature PCL suffered approximately 20% of degradation in simulated compost soil in 120 days. Ionizing radiation induced degradation rate increase with increasing radiation dose in the dose range studied, Fig.12, achieving 55% of degradation in the same degradation period. Probably it was because of aerobic condition of simulated compost soil test is performed.

PLLA suffered approximately 35% of degradation in the same period as PCL, Fig.13, and irradiation process promoted degradation rate increase with increasing radiation dose, achieving 70% of degradation in 120 days.

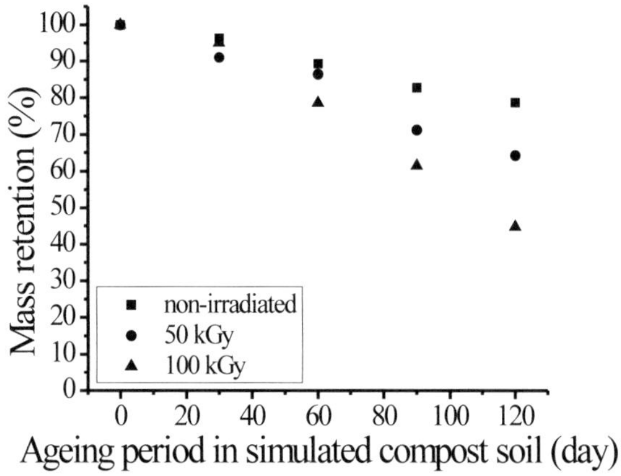

Figure 12. Mass retention variation versus biodegradation period in simulated compost soil of PCL samples non-irradiated and EB irradiated with radiation doses of 50 kGy and 100 kGy

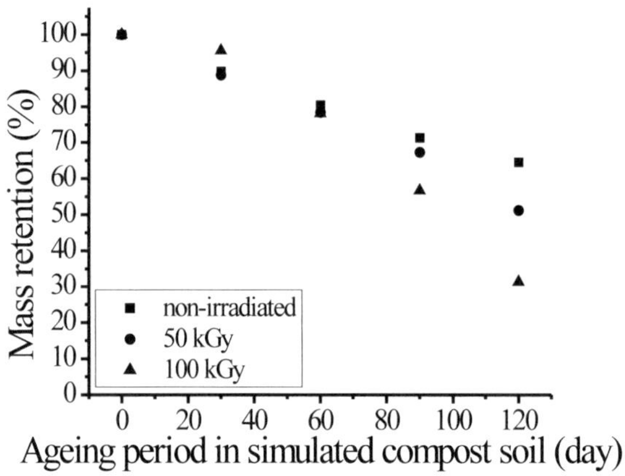

Figure 13. Mass retention variation versus biodegradation period in simulated compost soil, non-irradiated and EB irradiated PLLA samples with radiation doses of 50 kGy and 100 kGy.

Maharana et al. [27] cited that ionizing radiation does not affect glass transition temperature T_g, melting temperature T_f neither hydrolytic degradation of aliphatic polyesters. However, in our study it was possible to observe slight increase of biodegradation rate in simulated compost soil after 60 days. Probably this behavior is related to microorganism presence in soli that would favor the degradation process by produced oligomer consumption. According to the authors, as radiation induced reactions occur mainly in amorphous regions of polymers, it is important to know their crystallinity degree. Biodegradation is also affected by solid state morphology, primary chemical structure, for instance, functional groups existence and hydrophicity and

http://dx.doi.org/10.5772/56231

hydrophobicity equilibrium of PLA. Crystallinity degree is one of the main factors that controls degradation rate of solid polymers. In general, main chain scission occurs at esters bonds sites, leading to oligomer formation, which number after chain scission depends on the quantity of ester bonds present on PLA.

Normally, biodegradation occurs in three steps. In the first step, depolymerization occurs, then, in the second step depolymerized PLA produces lactic acid. Finally, lactic acid is consumed in citric acid cycle where it is transformed into CO_2 and H_2O in the presence of an enzyme produced by microorganism. PCL:PLLA 20:80 (w:w) blend suffers degradation of approximately 30% in 120 days, PCL slightly affected PLLA degradation in the blend, Fig.14.

Radiation absorbed dose of 50kGy did not affect significant effect of degradation rate, irradiated samples with 100kGy suffered few significant increase of degradation rate after 60 days.

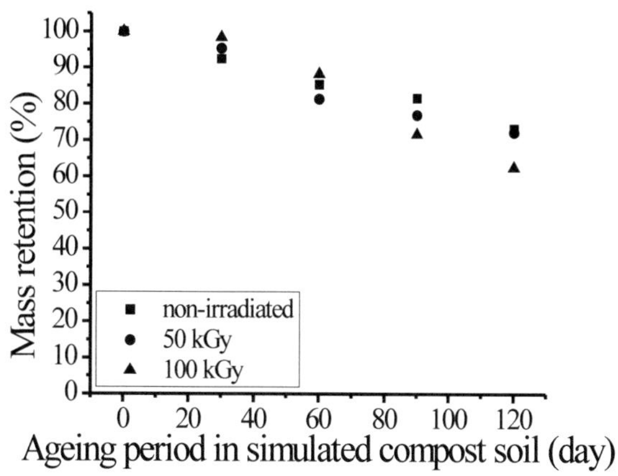

Figure 14. Mass retention variation versus biodegradation period in simulated compost soil of PCL:PLLA 20:80 (w:w) non-irradiated and EB irradiated with radiation doses of 50 kGy and 100 kGy

Absorbed radiation dose of 50 kGy did not significantly affect degradation of composite containing 5% of non-chemically treated coconut fiber, neither degradation rate in simulated compost soil, FIG.15. Samples irradiated with 100kGy suffered discrete increase of degradation rate after 60 days of test and over 120 days biodegradation tend to stabilize.

Absorbed radiation dose did not affect significantly biodegradation neither degradation rate of studied samples of composites containing 10% of non-chemically treated fibers, Fig.16.

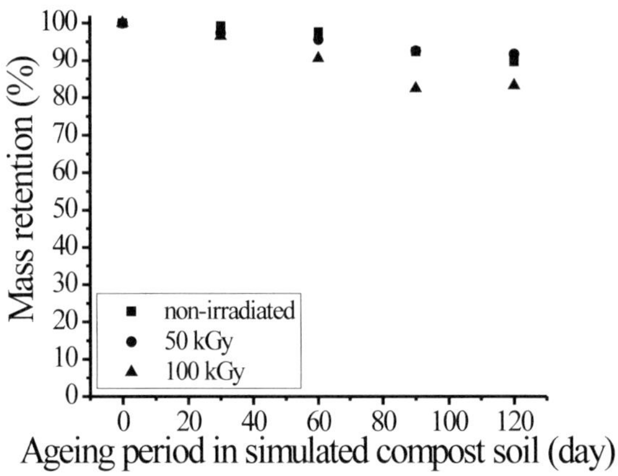

Figure 15. Mass retention variation versus biodegradation period in simulated compost soil of composites samples containing 5% of non-chemically treated coconut fibers, non-irradiated and EB irradiated with absorbed doses of 50 kGy and 100 kGy

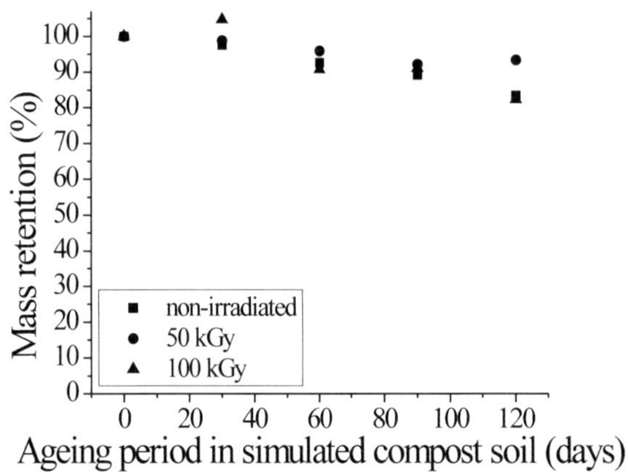

Figure 16. Mass retention variation versus biodegradation period in simulated compost soil of composites samples containing 10% of non-chemically treated coconut fibers, non-irradiated and EB irradiated with absorbed doses of 50 kGy and 100 kGy

It was observed during preparation of composites with acetylated coconut fibers that some kind of chemical reaction occurred during extrusion in some few events. Probably some vestiges of chemicals used for acetylation had remained on the coconut fibers. This fact could have affected degradation test of some samples of composites containing acetylated coconut fibers that started to present fissures favoring degradation on these points. Lucas et al. [31] cited that bio deterioration of thermoplastics occurs via two different mechanisms, erosion at surface and in the bulk. In the case of bulk erosion, fragments of total mass of polymer are lost and its molecular weight is altered because of bond rupture. This rupture is provoked by chemicals (H_2O, acids, bases, transition metal and radicals) or by radiation, however not by enzymes. They are very big to penetrate through bulk structure. Whereas in the case of surface erosion, matter is lost, though molecular weight of polymeric matrix does not alter. If chemical substances diffusion through the material is faster than bond scission of polymer, polymer suffers erosion. If the opposite occurs, process occurs mainly on the surface of polymeric matrix.

Radiation dose did not affect significantly biodegradation of composites containing 5% of acetylated fibers up to 60 days of test. Irradiated samples with 100kGy presented slight increase on the degradation rate after 90 days, Fig. 17.

On Fig. 18 it is possible to observe that composites samples containing 10% of acetylated fiber did not suffer significant alteration of degradation rate with radiation dose increase.

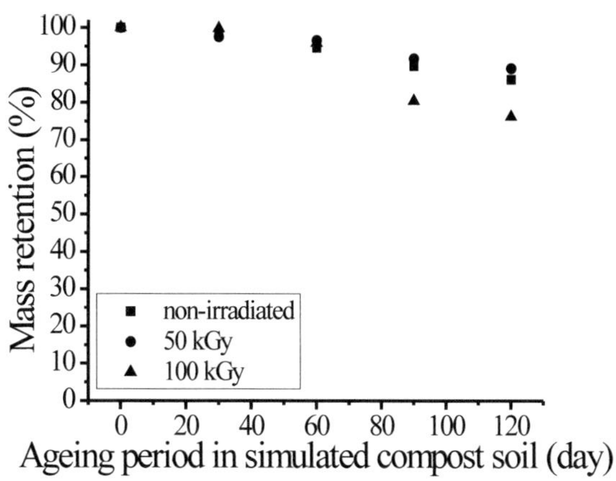

Figure 17. Mass retention variation versus period of biodegradation in simulated compost soil of composites samples containing 5% of acetylated fibers, non-irradiated and EB irradiated with absorbed doses of 50 kGy and 100 kGy

Higher degradation values found of those samples compared to non-chemically treated fibers could be related to the effect of fissures observed on the polymeric matrix during test that probably favored microbiological attack.

Mass retention results deviations were in average 7%, probably due to weight variation of sample to sample.

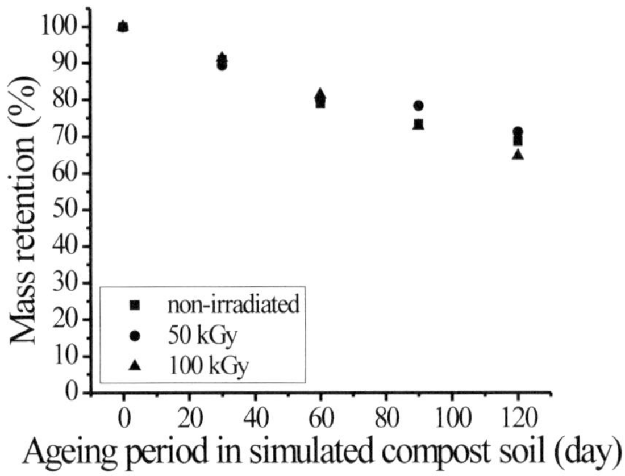

Figure 18. Mass retention variation versus period of biodegradation in simulated compost soil of composites samples containing 10% of acetylated fibers, non-irradiated and EB irradiated with radiation doses of 50 kGy and 100 kGy

4.3. Hydrolytic degradation

This section will present some aspects of hydrolytic degradation because PCL and PLLA homopolymers studied here are biomaterials.

Biodegradability of polymeric materials occurs in several steps. Initially, digestible macromolecules, that form polymeric chain, suffer enzymatic scission. This is followed by metabolism of scission parts, leading to progressive enzymatic degradation of macromolecules from chain ends. Instead, macromolecular oxidative cleavage occurs, inducing fragments metabolism. Anyway, chain fragments become small enough to be converted by microorganisms [13,20]. Enzymes are catalytic proteins that decrease activation energy of molecules favoring chemical reactions. Those proteins have large diversity and marked specificity, but are easily denatured by heating, radiation, surfactants, among others [31]. In Fig. 19 general mechanism of biodegradation of polymeric materials is presented.

According to Liu et al. [18], hydrolytic degradation of PCL and PLLA has been studied extensively. PLLA artifacts degradation is faster in the inner part than in the surface due to autocatalytic effect of carboxyl end groups. In the case of PCL, hydrolytic degradation is very low because of hydrophobicity and crystallinity. Authors reported that, in presence of proteinase K, PLLA degraded preferably at L-lactil units. Furthermore, enzymatic degradation occurred preferably on amorphous region of semi-crystalline PLLA polymers [11,18,19].

According to Lenglet et al. [19], hydrolytic degradation is a mass phenomenon and polyesters degradation with high size is auto catalyzed by carboxyl end groups initially present, or generated by ester bond cleavage. The three most important discoveries about polyester degradation performed in the last decade were about faster degradation in the inner portion of the sample and that degradation induces morphology and composition alteration. On the other hand, enzymes are macromolecules and cannot penetrate in a solid material. Then, enzymatic degradation occurs in two steps: adhesion of enzyme on the surface of sample followed by scission of polymeric chain catalyzed by enzyme that generally results in small alterations of properties of polymeric matrix. According those authors, highly crystalline PCL can be fully degraded in a couple of days in presence of *Pseudomonas* lipase, while hydrolytic degradation can take several years in 37°C (average temperature of human body). Kulkarni et al. [32] have cited that *Pseudomonas cepacia* lipase accelerates significantly PCL degradation. Interface activation of enzymes lipase type results mainly in conformational alteration of enzymes. Reaching substrate surface, they expose their active site and provide hydrophobic surface to the interaction with substrate molecular chains. The authors cited that several publications deal with the fundamentals of the theory of hydrolytic degradation and erosion of solid polymers. The basic modes, the surface erosion and the bulk degradation, depend on the relation between the rate of water/enzyme diffusion into the polymer, the rate of chain cleavage by water ions/enzymes, and the rate of transportation of scission products out of the solid. The rate of water diffusion into a polymer solid is strongly influenced by a number of structural parameters, its porosity, the crystallinity, the surface roughness, the hydrophobicity and the size of the sample. Most authors treat the enzymatic degradation of polymer solids exclusively as surface process. For hydrophilic enzymes it is usually considered to be difficult to penetrate into a hydrophobic polymer.

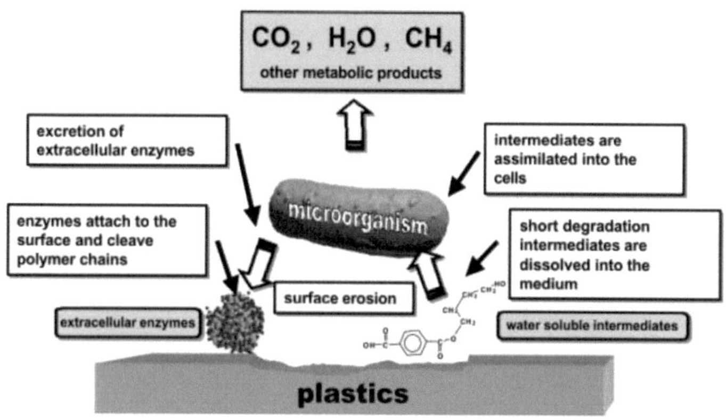

Figure 19. General mechanism of biodegradation of polymeric materials [13].

Loo et al. [11] cited that the rate of hydrolytic degradation for biopolymers like PGLA and PLLA is controlled by altering their physical properties; such as their molecular weights, degree of crystallinity and glass transition temperature (T_g). As mentioned previously, radiation has been known to alter the physical properties of polymers through main-chain scission and cross-linking. Semi-crystalline polymers, such as PLLA, are nonhomogeneous with a two-phase system consisting of amorphous and crystalline regions. During irradiation, energy is deposited uniformly and radicals are formed throughout the polymer in both the amorphous and crystalline regions. However, crystalline regions consist of chains that were more oriented and closely packed compared to the more open amorphous regions. As a result, oxygen, stabilizers and specific active radical species are excluded from the crystalline phase, and the irradiation chemical reaction paths in the amorphous and crystalline phases will therefore be different. According to Loo et al. [11], due to the close packing of the crystalline structure, the poor diffusion of oxygen into the crystalline region limits the formation of peroxyl free radicals and thus, the extent of chain scission. The ''cage effect'' also encourages the recombination of free radicals in the crystalline region. These factors play an important role in reducing the extent of e-beam degradation in PLLA.

5. Conclusion

Results of degradability test, enzymatic and in simulated compost soil, indicate that studied materials suffered accentuated degradation in enzymes presence and are not affected by negatively by radiation processing. Even though coconut fibers addition had slightly reduced degradation process, composites keep degrading through time. Artifacts produced utilizing the studied materials can be processed by ionizing radiation up to 100 kGy radiation doses without detriment of their biodegradability.

Acknowledgements

We are grateful to the financial support from JICA and IAEA. Additionally, to Dr. Akihiro Oishi and Dr. Kazuo Nakayama, from National Institute of Advanced Industrial Science and Technology – AIST, Japan, for samples preparation and valuable discussion; to Dr. Naotsugu Nagasawa and Dr. Masao Tamada, from Japan Atomic Energy Agency – JAEA, Japan, for samples preparation and irradiation. We also would like to thank Dr. Morsyleide Freitas Rosa from Embrapa for providing coconut fiber; to Prof. Dr. Hélio Wiebeck, and Mr. Wilson Maia from Laboratório de Matérias-Primas Particuladas e Sólidos Não Metálicos – LMPSol, Departamento de Engenharia de Materiais, Escola Politécnica da USP (EPUSP) for coconut fiber size reduction and segregation; also to Eng. Elisabeth S.R. Somessari, Eng. Carlos G. da Silveira, and Mr. Paulo de Souza Santos, from IPEN, for blends and composites irradiation. Furthermore, we would like to acknowledge Prof. Dr. Derval dos Santos Rosa, UFABC, Centro de Engenharia, Modelagem e Ciências Sociais Aplicadas (CECS), for degradability tests and valuable discussion. We would like also to thank Dr. Luci Diva Brocardo Machado and Eng. Marcelo Augusto Gonçalves Bardi for helpful discussion.

Author details

Yasko Kodama

Address all correspondence to: ykodama@ipen.br

Nuclear and Energy Research Institute- IPEN–CNEN/SP, Radiation Technology Center, São Paulo, Brazil

References

[1] Tsuji, H, & Ikada, Y. Blends of aliphatic polyesters. I. Physical properties and morphologies of solution-cast blends from poly (DL-lactide) and poly(ε-caprolactone). Journal of Applied Polymer Science (1996). , 60(1), 2367-2375.

[2] Kammer, H. W, & Kummerlowe, C. (1994). Poly (ε-caprolactone) Comprising Blends-Phase Behavior and Thermal Properties, *in* Finlayson, K. (ed.) *Advances in Polymer Blends and Alloys Technology*, Technomicv, USA, 5, , 132-160.

[3] Dell'Erba RGroeninckx G., Maglio G., Malinconico M., Migliozzi A. Imiscible polymer blends of semicrystalline biocompatible components: thermal properties and phase morphology analysis of PLLA/PCL blends. Polymer (2001 4). , 2001(42), 1-7831.

[4] Yoshii, F, Darvis, D, Mitomo, H, & Makuuchi, K. Crosslinking of poly (ε-caprolactone) by radiation technique and its biodegradability. Radiation Physics and Chemistry (2000 5). , 2000(57), 1-417.

[5] Zhang, J, Duan, Y, Sato, H, Tsuji, H, Noda, I, Yan, S, & Ozaki, Y. Crystal modifications and thermal behavior of poly (L-lactic acid) revealed by infrared spectroscopy. Macromolecules (2005 3). , 2005(38), 1-8012.

[6] Mochizuki, M, & Hirami, M. Structural effects on the biodegradation of aliphatic polyesters. Polymers for AdvancedTechnology (1997 8). , 1997(8), 1-203.

[7] Michler, G. H. ed) ((2008). Electron Microscopy of Polymers, Springer-Verlag.

[8] Sawyer, L. C, Grubb, D. T, & Meyers, G. F. eds) ((2008). Polymer Microscopy 3rd ed, Springer.

[9] Azevedo, H. S, & Reis, R. L. (2005). Understanding the enzymatic degradation of biodegradable polymers and strategies to control their degradation rate. In: Reis RL, Román JS (eds). Biodegradable Systems in Tissue Engineering and Regenerative Medicine. CRC, Boca Raton, , 177-197.

[10] Chmielewski, A. G. New Trends in radiation processing of polymers, In: International Nuclear Atlantic Conference; Encontro Nacional de Aplicações Nucleares, 7th, aug. sept. 2, 2005, Santos, SP. Anais... São Paulo: ABEN, (2005). , 28.

[11] Loo, J. S. C, Ooi, C. P, & Boey, F. Y. C. Degradation of poly(lactide-co-glycolide) (PLGA) and poly(l-lactide) (PLLA) by electron beam radiation. Biomaterials (2005). , 2005(26), 1-1359.

[12] Kantoglu, Ö, & Güven, O. Radiation induced crystallinity damage in poly(L-lactic acid), Nuclear Instruments and Methods in Physical Research B (2002 1). , 2002(19), 1-259.

[13] Müller, R. J. ((2005). Biodegradability of Polymers: Regulations and Methods for Testing *in*: STEINBÜCHEL, A. (ed.), Biopolymers- General Aspects and Special Applications, Wiley-VCH Verlag GmbH & Co. KgaA, n. 19, , 10, 365-374.

[14] Kuo, S-W. Hydrogen-bonding in polymer blends. Journal of Polymer Research (2008 1). , 2008(15), 1-459.

[15] Tsuji, H, & Ishizaka, T. Blends of aliphatic polyesters, VI. Lipase-catalyzed hydrolysis and visualized phase strucuture of biodegradable blends from poly(e-caprolactone) and poly(L-lactide). International Journal of Biological Macromolecules (2001 2). , 2001(29), 1-83.

[16] Kikkawa, Y, Suzuki, T, Tsuge, T, Kanesato, M, Doi, Y, & Abe, H. Phase structure and enzymatic degradation of poly(L-lactide)/atactic poly(3-hydroxybutyrate) blends: an atomic force microscopy study. Biomacromolecules (2006). , 2006(7), 1-1921.

[17] Nishino, T, & Hirao, K. Kotera M. X-ray diffraction studies on stress transfer of kenaf reinforced poly(L-lactic acid) composite. Composites: Part A (2006). , 2006(37), 1-2269.

[18] Liu, L, Li, S, Garreau, H, & Vert, M. Selective Enzymatic Degradations of Poly(L-lactide) and Poly(ε-caprolactone) Blend Films. Biomacromolecules (2000). , 2000(1), 1-350.

[19] Lenglet, S, Li, S, & Vert, M. Lipase-catalysed degradation of copolymers prepared from e-caprolactone and DL-lactide. Polymer Degradation and Stability (2009 9). , 2009(94), 1-688.

[20] Kolybaba, M, Tabil, L. G, Panigrahi, S, Crerar, W. J, Powell, T, & Wang, B. (2003). Biodegradable polymers: past, present and future. 2003 CSAE/ASAE Annual Intercectional Meeting Sponsered by the Red Rive Section of ASAE. Fargo, North Dakota, USA, October , 3-4.

[21] Calado, V, Barreto, D. W, & Almeida, D. J.R.M. The effect of a chemical treatment on the structure and morphology of coir fibers. Journal of Materials Science Letters (2000). , 2000(19), 1-2151.

[22] Almeida, D, Calado, A. L. F. S, & Barreto, V. D.W. Acetilação da fibra de bucha (Luffa cylindrica) Polímeros: Ciência e Tecnologia (2005 1). , 2005(15), 1-59.

[23] Spinks, J. W. T, & Woods, R. J. An Introduction to Radiation Chemistry. 3rd ed. USA: John Wiley and Sons; (1990).

[24] Calil, M. R, Gaboardi, F, Bardi, M. A. G, Rezende, M. L, & Rosa, D. S. Enzymatic degradation of poly(ε-caprolactone) and cellulose acetate blends by lipase and α-amilase. Polymer Testing (2007 2). , 2007(26), 1-257.

[25] Sivalingam, G, Vijayalakshmi, S. P, & Madras, G. Enzymatic and thermal degradation of poly(e-caprolactone), poly(D,L-lactide, and their blends. Industrial & Engineering Chemistry Research (2004 4)., 2004(43), 1-7702.

[26] Cottam, E, Hukins, D. W. L, Lee, K, Hewitt, C, & Jenkins, M. J. Effect of sterilization by gamma irradiation on the ability of polycaprolactone (PCL) to act as a scaffold material. Medical Engineering&Physics (2009 3)., 2009(31), 1-221.

[27] Maharana, T, Mohanty, B, & Negi, Y. S. Melt-solid polycondensation of lactic acid and its biodegradability. Progress in Polymer Science (2009 3)., 2009(34), 1-99.

[28] Salazar, V. L. P, & Leão, A. L. Biodegradação das fibras de coco e de sisal aplicadas na indústria automotiva. Revista Energia na Agricultura (2006)., 2006(21), 2-99.

[29] Alauzet, N, Roussos, S, Garreau, H, & Vert, M. Microflora dynamics in earthworms casts in an artificial soil (biosynthesol) containing lactic acid oligomers Brazilian archives of biology and technology (2001 4)., 2001(44), 2-113.

[30] Lotto, N. T, Calil, M. R, Guedes, C. G. F, & Rosa, D. S. The effect of temperature on the biodegradation test. Materials Science and Engineering C (2004 2)., 2004(24), 1-659.

[31] Lucas, N, Bienaime, C, Belloy, C, Queneudec, M, Silvestre, F, & Nava-saucedo, J. E. Polymer biodegradation: mechanisms and estimation techniques. Chemosphere (2008 7)., 2008(73), 1-429.

[32] Kulkarni, A, Reiche, J, Hartmann, J, Kratz, K, & Lendlein, A. Selective enzymatic degradation of poly(e-caprolactone) containing multiblock copolymers. European Journal of pharmaceuticals and biopharmaceutics (2008 6)., 2008(68), 1-46.

Chapter 7

Intradiol Dioxygenases — The Key Enzymes in Xenobiotics Degradation

Urszula Guzik, Katarzyna Hupert-Kocurek and Danuta Wojcieszyńska

Additional information is available at the end of the chapter

http://dx.doi.org/10.5772/56205

1. Introduction

Aromatic compounds are derived from both natural and anthropogenic sources. Under natural conditions, arenes are formed as a result of the pyrolysis of organic materials at high temperatures during forest, steppe and peatland fires, and during volcanic eruptions. Biogenic aromatic compounds like aromatic amino acids and lignin, the second most abundant organic compound in the environment, are universally distributed in nature. Many species of plants, especially willow (*Salix*), thyme (*Thymus vulgaris*), camomile (*Chamomilla recutita*), bean (*Phaesoli vulgaris*) or strawberry (*Fregaria ananasa*), water plants as sweet flag (*Acorus calamus*) and many species of alga are known to produce aromatic compounds as secondary metabolites [1-4]. A lot of aromatic compounds are introduced to the environment as contaminating compounds from chemical, pharmaceutical, explosive, dyes, and agrochemicals industry. Chloro-, amino- and nitroaromatic derivatives, biphenyls, polycyclic aromatic hydrocarbons accumulate in the soil and water. They are toxic to living systems including humans, animals, and plants. Moreover, most of them may bioaccumulate in the food chain and have mutagenic or carcinogenic activity [5-8].

Due to the delocalization of π orbitals aromatic compounds are very stable [3]. For many years it has been searching for microorganism able to breakdown these kind of substrates. Among the bacteria, representatives of the genus *Pseudomonas, Sagittula, Streptomyces, Agrobacterium, Acinetobacter, Arthrobacter, Burkholderia, Bacillus, Stenotrophomonas, Brevibacterium, Sphingomonas, Geobacillus, Rhodococcus, Nocardia, Corynebacterium, Alcaligenes, Gordonia*, or *Citrobacter* are active in this respect [9-22].

The strategy for degradation of aromatic structure comprises hydroxylation and cleavage of the aromatic ring with the usage of oxygenases [8,23,24]. These reactions are initiated by electrophilic attack of molecular oxygen and they are retarded by the presence of electron-withdrawing substituents. For this reason the velocity of reaction is connected with the kind of substituents in the aromatic ring. Substituents of the first group such as: $-NH_2$, -OH, -OCH_3, $-CH_3$, $-C_6H_5$ activate aromatic ring for electrophilic attack. In contrary, the substituents of the second group ($-NO_2$, $-N(CH_3)_3^+$, -CN, -COOH, -COOR, $-SO_3H$, -CHO, -COR) and the third group (-F, -Cl, -Br, -I) impede the electrophilic attack by dioxygenases [3,25-28].

Hydroxylation into the dihydroxylated intermediates, the first step in the oxidative degradation of aromatic compounds, is catalyzed by monooxygenases or hydroxylating dioxygenases (Fig. 1). Aromatic monooxygenases are divided into two groups: activated- ring monooxygenases and nonactivated- ring enzymes and usually contain FAD as a prosthetic group. These enzymes posses a dinuclear iron centre with two oxo- bridged iron atoms (Fe-O-Fe) in the active site [36-39]. Aromatic monooxygenases activate molecular oxygen through the formation of a reactive flavin (hydro)peroxide which can attack the substrate [31,33-35].

Aromatic ring hydroxylating dioxygenases that belong to the Rieske non-heme iron oxygenases are multicomponent enzyme systems. The reaction catalyzed by hydroxylating dioxygenases requires two electrons from a NAD(P)H that are consecutively transferred to the terminal oxygenase component through a electron carries such as ferredoxin and/or a reductase [36].

As a result of hydroxylation key intermediates such as catechol, protocatechuic acid, hydroxyquinol, or gentisic acid are formed. These products are substrates for ring-cleaving dioxygenases [17,37-39]. The ring-cleaving dioxygenases couple O_2 bond cleavage with ring fission of hydroxylated derivatives either between the two hydroxyl group (*ortho* cleavage) or beside one of these (*meta* cleavage) [23,40].

Intradiol dioxygenases catalyze the intradiol cleavage of the aromatic ring at 1,2-position of catechol or its derivatives (protocatechuic acid, hydroxyquinol) with incorporation of two atoms of molecular oxygen into the substrate. It leads to the production of *cis,cis*-muconic acid or its derivatives (3-carboxy-*cis,cis* muconic acid, 3-hydroxy-*cis,cis* muconic acid). The *cis,cis*-muconic acid is then subsequently transformed by a muconate cycloisomerase to muconolactone. Muconolactone isomerase shifts the double bond to form 3-oxoadipate-enol-lactone, the first common intermediate of the catechol and protocatechuate or hydroxyquinol branch (Fig. 1) [38,41,42].

Extradiol catechol and protocatechuate dioxygenases catalyze the ring fission between position C2 and C3 of catechol ring and between position C2 and C4 or C4 and C5 of protocatechuate ring respectively. Products of these reactions (2-hydroxymuconic semialdehyde or carboxy-2-hydroxymuconic semialdehyde) are transformed finally to pyruvic acid and acetaldehyde in catechol pathway and pyruvic acid and acetaldehyde or pyruvic acid and oxaloacetic acid in protocatechuate pathway (Fig. 1) [37,43].

Degradation of aromatic compounds by the *meta* pathway often leads to more toxic intermediates such as acyl chloride that is why in the degradation of some arenes the *ortho* cleavage is preferred [5,6]. Recently intradiol dioxygenases with extremely high activity were isolated.

http://dx.doi.org/10.5772/56205

They could be applicable not only in biodegradation but also in industry processes since products of the *ortho* cleavage are valuable intermediates in chemical synthesis [44,45].

Because some aromatics have been present in the environment for a very short time, bacteria have not evolved efficient pathways for their degradation and for that reason the construction of such pathways may be required. Therefore, a detailed understanding of the molecular bases as well as enzymes of known pathways is essential [46].

Our chapter is divided into three parts: first we classify bacterial intradiol dioxygenases, key enzymes in aromatic compounds degradation, taking under consideration phylogenetic relationships among dioxygenases from various strains and their substrate specificity. Next, we present current knowledge about molecular structure of these dioxygenases and hypothetical mechanisms for aromatic ring cleavage. And finally, we describe modifications increasing biodegradation potential of these enzymes.

2. Classification of bacterial intradiol dioxygenases

Many studies on intradiol dioxygenases concentrate on the understanding of their biochemical and structural properties. Based on the results of these studies and taking into consideration substrate specificity of the *ortho* fission dioxygenases they were divided by Vetting and Ohlendorf [47] into two families: catechol 1,2-dioxygenases and protocatechuate 3,4-dioxygenases, including hydroxyquinol 1,2-dioxygenases. In turn, Hammer et al. [12] and Contzen and Stolz [48] divided protocatechuate 3,4-dioxygenases into two types. Type I dioxygenases catalyze ring fission of protocatechuic acid and its derivatives while type II enzymes cleave *ortho* diphenols carrying a larger substituent at the 4-position of the aromatic ring [12,48].

Although Murakami et al. [14] observed partial activity of hydroxyquinol 1,2-dioxygenase towards catechol and Pandeeti and Siddavattam [49] isolated catechol 1,2-dioxygenase which was active strictly against catechol and 1,2,3-benzenotriol there is no reports about activity of catechol 1,2-dioxygenases towards hydroxyquinol [18,21,50,51]. Therefore, we propose to divide the intradiol dioxygenases into three classes: catechol 1,2-dioxygenases, protocatechuate 3,4-dioxygenases, and hydroxyquinol 1,2-dioxygenases. Representatives of these three classes are presented in Table 1. Based on the results of biochemical and genetic comparison of intradiol dioxygenases Perez-Pantoja et al. [52] distinguished hydroxyquinol 1,2-dioxygenases as a separate cluster that may support our suggestions.

The catechol 1,2-dioxygenases family can be divided into several subclasses depending on their substrate specificity. Initially this family was divided into two subclasses: type I dioxygenases with activity against catechol and lack or weak activity against chlorocatechol and type II dioxygenases, which cleave only chlorocatechols [53]. Then, based on their amino acid sequences Liu et al. [54] divided family of catechol 1,2-dioxygenases into three groups: catechol 1,2-dioxygenases from gram-negative bacteria, chlorocatechol 1,2-dioxygenases from gram-negative bacteria, and both catechol- and chlorocatechol 1,2-dioxygenases from gram-positive bacteria. However, both of these classifications do not take into consideration substrate specificity of these enzymes towards methylcatechols. High substrate specificity against these compounds was observed for intradiol dioxygenase isolated from genus *Rhodococcus* [55-57]. Above studies divided catechol dioxygenases into three subclasses. The first subclass comprises the enzymes with high enzymatic activity towards chlorocatechols (CCs), the second one includes enzymes with moderate activities for catechol only or catechol and 4-methylcatechol (4-MC). The third subclass consists of the enzymes with high activity for 3- and 4-methylcatechols (3-MC and 4-MC) [50,55].

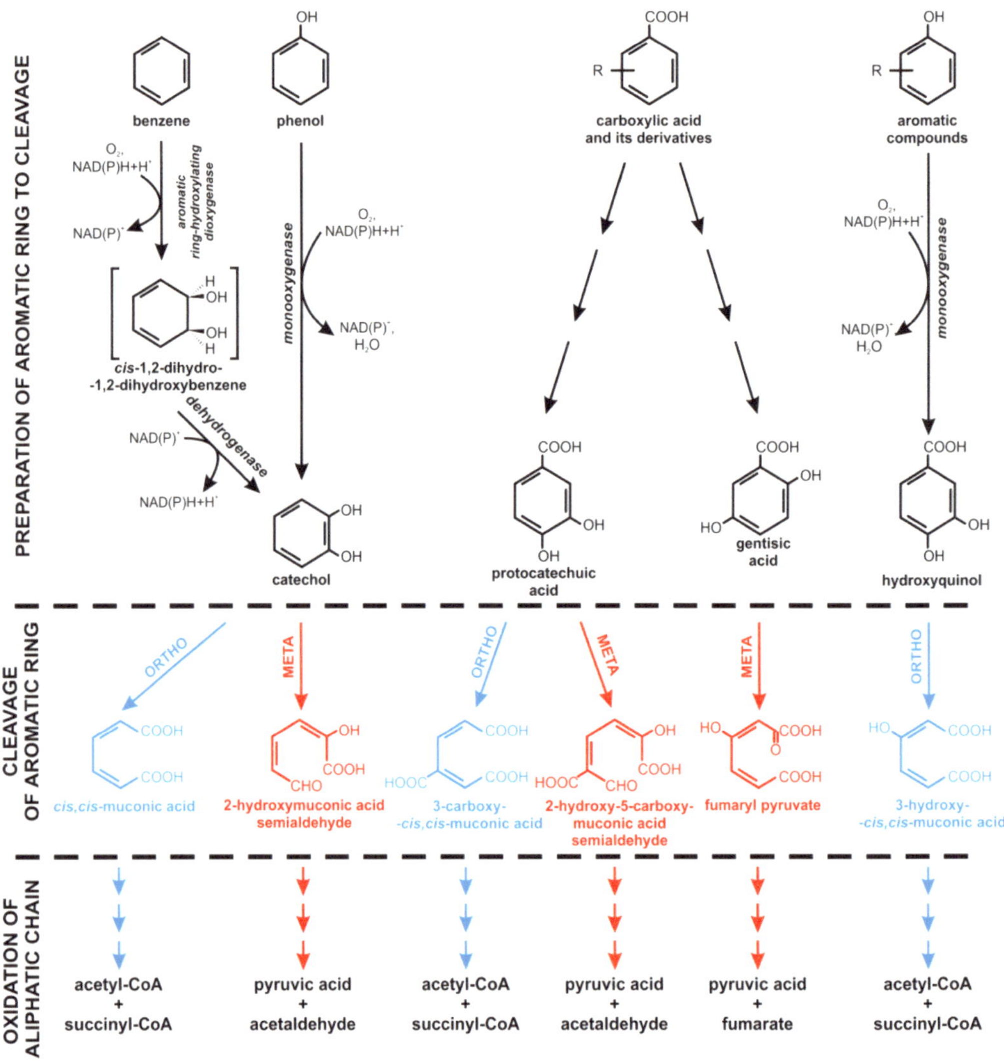

Figure 1. Pathways of aromatic compounds degradation [36,40,43].

Enzyme	Microorganism	Gene	AC Number	Substrates	References
Catechol 1,2-dioxygenase	*Acinetobacter calcoaceticus* ADP-96	-	-	catechol 3-isopropylcatechol 4-methylcatechol 3-methylcatechol	[58]
	Acinetobacter calcoaceticus ADP1	-	-	catechol	[47]
	Acinetobacter radioresistens	-	-	catechol 4-methylcatechol 3-methylcatechol	[59]
	Acinetobacter radioresistens S13	$catA_A$ $catA_B$	AF380158 AF182166	catechol 4-methylcatechol	[53,60]
	Acinetobacter sp. DS002	*catA*	-	catechol	[49]
	Alcaligenes eutrophus B.9	-	-	catechol 4-methylcatechol	[25]
	Alcaligenes eutrophus CH34	*catA*	NC_007973.1	catechol	[61]
	Arthrobacter BA-5-17	*catA-II*	AB109791.1	catechol 3-methylcatechol 4-methylcatechol	[62]
	Burkholderia sp. TH2	*catA1*	AB035483	catechol 3-methylcatechol 4-methylcatechol	[63]
	Burkholderia sp. TH2	*catA2*	AB035325	catechol 4-methylcatechol	[63]
	Citrobacter sp. SA01	-	-	catechol	[22]
	Gordonia polyisoprenivorans	-	-	catechol	[21]
	Pseudomonas aeruginosa TKU002	-	-	Catechol 4-methylactechol pyrogallol	[64]
	Pseudomonas putida N6	*catA*	EU000396.1	catechol	[38]
	Pseudomonas sp. B13	-	-	catechol 3-chlorocatechol 4-chlorocatechol 3-methylcatechol 4-methylcatechol	[25]
	Rhodococcus erythropolis AN-13	*catA*	D83237	catechol	[55]

Enzyme	Microorganism	Gene	AC Number	Substrates	References
	Rhodococcus opacus 1CP	*catA*	X99622.1	catechol 3-methylcatechol 4-methylcatechol	[50]
	Rhodococcus opacus 1CP	*clcA*	AF003948.1	4-chlorocatechol	[65]
	Rhodococcus opacus 1CP	*clcA2*	AJ439407.1	3-chlorocatechol	[65,66]
	Rhodococcus sp. NCIM 2891	-	-	catechol	[20]
	Sphingomonas xenophaga QYY	-	-	catechol	[67]
	Stenotrophomonas maltophilia KB2	*catA*	EU000397.1	catechol	[68]
Protocatechuate 3,4-dioxygenase	*Acinetobacter calcoaceticus*	*pcaHG*	-	protocatechuate	[10,69]
	Acinetobacter lwoffii K24	*pcaHG*	AY099487	protocatechuate	[70]
	Acinetobacter sp. ADP1	*pcaHG*	NC_005966.1	protocatechuate	[16]
	Agrobacterium radiobacter S2	*pcaH1G1* *pcaH2G2*	AF230649 AF230650	protocatechuate protocatechuate 4-sulfocatechol	[48]
	Agrobacterium radiobacter S2	-	-	protocatechuate4-sulfocatechol	[12]
	Brevibacterium fuscum	-	-	protocatechuate	[71]
	Burkholderia sp. NCIMB 10467	*pcaHG*	YP552570 YP552569	protocatechuate	[19]
	Geobacillus sp.	-	-	protocatechuate	[18]
	Hydrogenophaga palleronii S1	-	-	protocatechuate 4-sulfocatechol 3,4-dihydroxyphenyl-acetate 3,4-dihydroxy-cinnamate	[12]
	Moraxella sp.GU2	-	-	protocatechuate	[11]
	Pseudomonas cepacia	-	-	protocatechuate	[72-74]
	Pseudomonas cepacia DBO1	-	-	protocatechuate	[75]

Enzyme	Microorganism	Gene	AC Number	Substrates	References
	Pseudomonas fluorescens A.3.12	-	-	protocatechuate	[76]
	Pseudomonas marginata	*pcaHG*	U33634	protocatechuate	[77]
	Pseudomonas putida ATCC23975	*pcaHG*	L14836.1	protocatechuate	[78-80]
	Pseudomonas sp. HR199	*pcaHG*	Y18527	protocatechuate	[81]
	Rhizobium trifolii	-	-	protocatechuate	[82]
	Stenotrophomonas maltophilia KB2	*pcaH* *pcaG*	JQ364963 JQ364964	protocatechuate	[68]
	Streptomyces sp. 2065	*pcaHG*	AF109386.2	protocatechuate	[15]
Hydroxyquinol 1,2-dioxygenase	*Arthrobacter* sp. BA-5-17	*tftH*	AB016258	hydroxyquinol, catechol	[14]
	Azotobacter sp. GP1	-	-	hydroxyquinol, chlorohydroxyquinol	[41,83]
	Burkholderia cepacia AC1100	*tftH*	U19883.1	hydroxyquinol	[84]
	Burkholderia sp. AK-5	-	-	hydroxyquinol	[85]
	Nocardioides simplex 3E	*chqB*	AY822041.2	hydroxyquinol, 6-chlorohydroxyquinol, 5-chlorohydroxyquinol,	[17,86]
	Pseudomonas sp. 1-7	*pdcDE*	FJ821777.2	hydroxyquinol	[8]
	Pseudomonas sp. WBC-3	*pnpG*	EF577044.1	hydroxyquinol	[51]
	Ralstonia eutropha JMP134	*tcpC*	AF498371	6-chlorohydroxyquinol	[87]
	Ralstonia pickettii DTP0602	*hadC*	D86544	hydroxyquinol pyrogallol 3-methylcatechol 6-chlorohydroxyquinol	[88]
	Rhizobium sp. MTP-10005	*graB*	AB266211.1	hydroxyquinol	[89]
	Rhodococcus opacus SAO101	*npcC*	AB154422	hydroxyquinol	[90]

Table 1. Intradiol dioxygenases from various microorganisms

3. Molecular structure and reaction mechanism of intradiol dioxygenases

Intradiol dioxygenases belong to metalloenzymes with non-heme iron (III) that plays a crucial role for the binding and activation of dioxygen in the cleavage reaction [55,59]. Metal ion is ligated by two tyrosine and two histidine residues that are highly conserved in all investigated intradiol dioxygenases. The fifth ligand is a hydroxyl group derived from the solvent.

Two structurally different intradiol dioxygenases families can be singled out: catechol/hydroxyquinol 1,2-dioxygenases and protocatechuate 3,4-dioxygenases. Catechol/hydroxyquinol 1,2-dioxygenases are dimmers $(\alpha Fe^{3+})_2$ with identical or similar subunits while protocatechuate 3,4-dioxygenases are typically composed of two homologous subunits in oligomeric complexes $(\alpha\beta Fe^{3+})_{2-12}$ [66] (Fig. 3).

3.1. Catechol 1,2-dioxygenases characterization

Catechol 1,2-dioxygenases catalyze cleavage of catechol and its derivatives. The tertiary structure of the enzymes from this class is similar [53,59,61,64,66]. They form homodimers with two subunits connected with helical zipper motif. Each subunit has a molecular mass of 29.0-38.6 kDa.

Subunits of catechol 1,2-dioxygenases contain two molecules of phospholipids bound into a hydrophobic cavity at the dimeric interface [38,47,49,50,53,59,61,64,66,67]. On the base of electron density and the stereochemistry these phospholipids were identified as a phosphatidylcholine with two C14-15/C12-C17 hydrophobic tails [66]. It is suggested that the phospholipid chains may modulate the activity of catechol 1,2-dioxygenases by increasing the local concentration of substrate or the lipids act as an effector molecule. The binding of phospholipid to hydrophobic amino acid of the active site can lead to a conformational change at this place. During catalysis catechol is cleaved to *cis,cis-* muconic acid which is known to be toxic to the cell at low levels. Perhaps the next enzyme of intradiol pathway- *cis-cis-* muconate lactonizing enzyme, interacts with the hydrophobic tunnel and allows the proteins to pass the product from one enzyme to next. Moreover, phospholipids can alter the phospholipid bilayer in response to toxic levels of aromatic compounds [47,66].

Unique helical N-terminal domain was first described by Vetting and Ohlendorf [47] for catechol 1,2-dioxygenase from *Acinetobacter calcoaceticus* ADP1. Sequence aligments indicate that this helical domain is conserved among all members of catechol 1,2-dioxygenase class.

Catechol 1,2-dioxygenase from *Acinetobacter calcoaceticus* ADP1 is an elongated molecule, which is composed of helices and β sheets. The overall shape of this enzyme is similar to boomerang and it is divided into three domains: the two catalytic domains at each end of the molecule, and one central linker domain which is composed of several helices. The catalytic domain consists of two mixed topology β sheets which form a β sheets sandwich [47]. The similar structure is also observed in another catechol 1,2-dioxygenases [38,50].

The linker domain of catechol 1,2-dioxygenases contains 12 α helices. Five of them belong to the N-terminal chain of each monomer and one extending from each catalytic domain. The

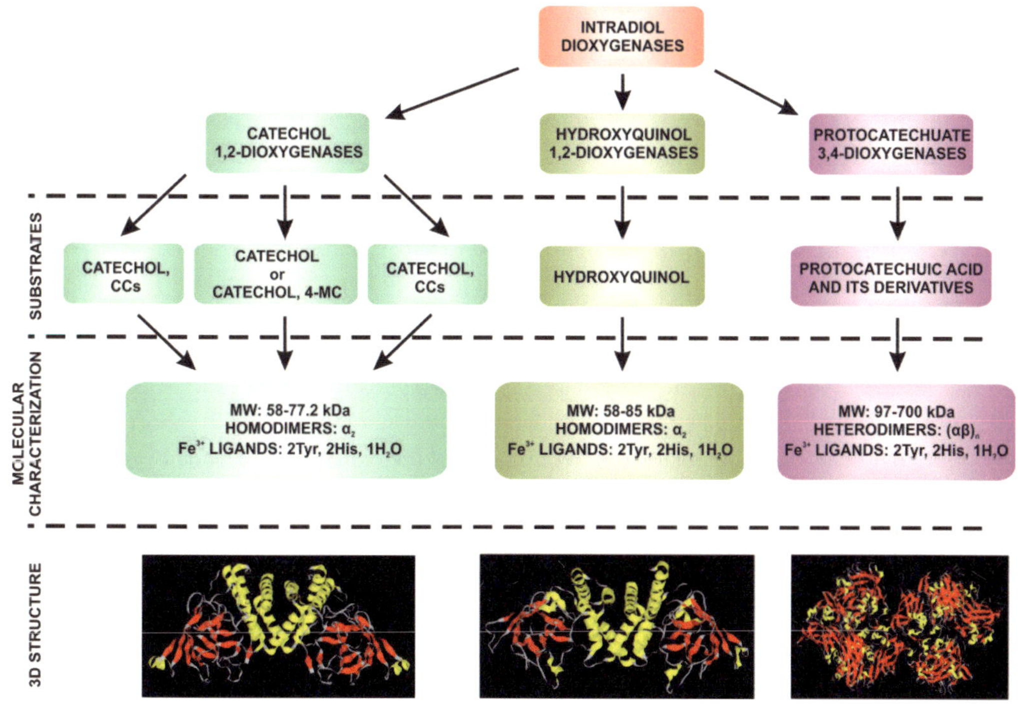

Figure 2. Characterization of intradiol dioxygenases

active site with ferric ion is located between β sheets sandwich and the linker domain [47]. Matera et. al. [50] compared structure of catechol 1,2-dioxygenase from Gram- positive strain *Rhodococcus opacus* 1CP with the enzyme from *Acinetobacter calcoaceticus* described above and observed that the differences in the structure are localized at the N-terminus. In this terminal chain helix 5 is shorter, strand 1 and helix 7 are missing and a new short helix is present between position 124 and 128. Moreover the C-terminal chain of catechol 1,2-dioxygenase from Gram-positive strain is shorter than in the other catechol 1,2-dioxygenases from Gram-negative bacteria.

As it was mentioned earlier the metal ion is ligated by four amino acid residues and a hydroxyl group from water in trigonal bipyramidal geometry. One molecule of tyrosine and one histidine are the axial ligands, whereas the second Tyr, His and a solvent molecule make up the equatorial plane. [38,47,50,53]. Since the two Fe^{3+} ions from catalytic domains are separated by more than 40 Å Vetting and Ohlendorf [47] suggested that they act structurally and catalytically independently. It explains the lack of allosteric effect [25,46,57,59,61,70,80]. However, catechol 1,2-dioxygenase from *Acinetobacter radioresistens* exhibits a sigmoidal kinetics indicating cooperation of subunits [59].

The substrate specificity of catechol 1,2-dioxygenases depends on electronic factors, which can play a significant role in determining the efficiency of conversion of catechol derivatives. Currently, there is little information about the influence of steric factors on catalytic properties of catechol 1,2-dioxygenases [50]. Sauret-Ignazi et al.[61] revealed that very high inhibition constant is connected with steric hindrance, for example by methoxyl group. However another studies have indicated that a steric hindrances are not determining the rate-limiting step but there is a strong correlation between substituent electronegativity and k_{cat} [25,50,61].

The catalytic cycle of catechol 1,2-dioxygenases involve binding of the catechol or its derivatives as a dianion with simultaneous dissociation of one tyrosine and water molecule (a-e). One hydroxyl group of substrate binds in an axial position *trans* to His and the other binds in an equatorial position *trans* to Tyr. The ligand position *trans* to equatorial His is unoccupied and is postulated to be the oxygen-binding pocket [28,47,50,61]. The active site geometry is converted into a square-pyramidal geometry [91]. The substrate chelates asymmetrically to the iron(III) center. The long Fe-O bonds is *trans* to a tyrosinate ligand and the short Fe-O bond is *trans* to a neutral histidine ligand [40,91]. Binding of the distal oxygen leads to species with a peroxide bridge between the ferric ion and catechol substrate [92]. At the initial stage dioxygen molecule is one-electron reduced by the catecholate ligand and in this way it formes superoxide radical anion that interacts with iron(III) (f,g). In the next step superoxide and catecholate radicals recombine forming peroxo species (h). The last can easily change its conformation because of the hydroxyl group that becomes a ketone one, which is a much weaker ligand. Arginine 457 is supposed to stabilize the carboanion produced by the C3-O bond ketonization [92]. In consequences a coordination site of the axial tyrosine is open and it allows returning of the tyrosine to the coordination sphere of iron. The tyrosine binds to iron (III) as a phenolate (i), whereas the proton either neutralizes the peroxide ligand or it is released from the active site [40,91,92].

In the first case, proton remains in the active site (i) and the peroxide oxygen atom is protonated. It prevents oxidation of iron to a high-valent Fe(IV)=O state during the O-O bond cleavage. As a result peroxide ligand undergoes a Criegee rearrangement with O-O bond heterolysis and migration of the acyl group to the peroxide oxygen (k). It is also possible that the peroxide ligand converts to muconic anhydrate product via O-O bond cleavage made easier by the equatorial tyrosine, which is oxidized to a tyrozyl radical (j). The reduction of tyrozyl radical to tyrosinate is connected with the oxyl radical attack on the carbonyl carbon [40].

In the second case, proton is released from the active site (n) and peroxide bridge is cleaved to reactive oxoferryl group and an alkoxyl radical (o). Alkoxyl radical attacks the adjacent carbonyl carbon, followed by the ring expansion and protonation of the oxo ligand leads to the cyclic anhydride (k). Hydrolysis of the anhydride leads to the final reaction product – muconic acid (l,m) (Fig. 3) [40,92].

The electrophilic attack of molecular oxygen on iron (III)-bound catecholate is judged to be an essential step in the catalytic reaction of catechol 1,2-dioxygenase [50]. For this type of chemical reaction the σ-complex is assumed to be next to the transition state. The formation of σ-complex requires the highest free energy of activation and must consequently be rate-limiting [25].

http://dx.doi.org/10.5772/56205

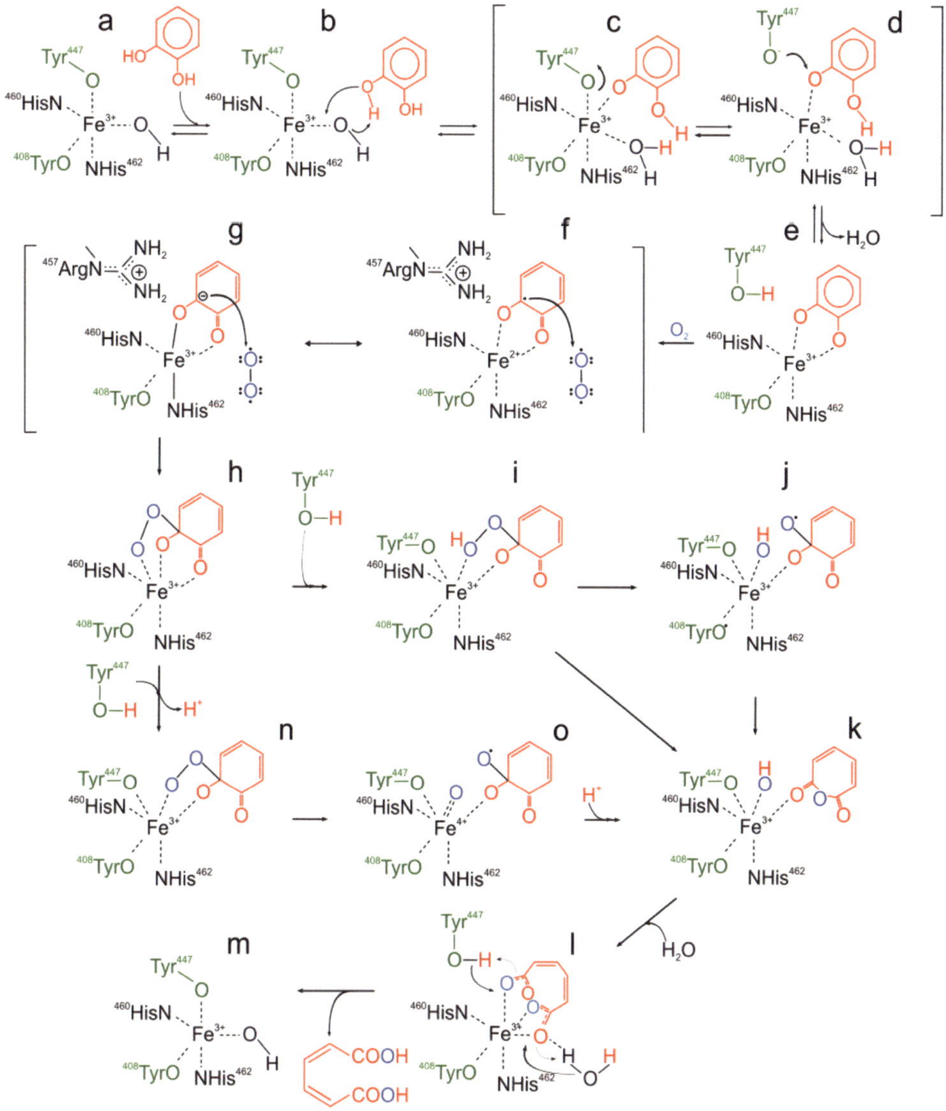

Figure 3. The catalytic cycle of the intradiol dioxygenases [92].

Sauret-Ignazi et al. [61] observed that monophenolic compounds are also bound by enzyme and they suggested that only one deprotonated hydroxyl may be sufficient for binding of the substrate to the active site.

Matera et al. [50] calculated parameter for nucleophilic reactivity for catechol, 3-chloro-, 4-chloro-, 3-methyl-, 4-methylcatechols, and pyrogallol. They revealed that the interaction between the frontier orbitals, the highest occupied molecular orbital (HOMO) of nucleophiles (catechols) and the lowest unoccupied molecular orbital (LUMO) of electrophile (the active site iron ion), is essential and lowers the activation barrier. Since the nuclephilic attack of the aromatic π electrons of the iron(III)-bound catecholate on the oxygen molecule is an important step in the cleavage reaction, the HOMO of the catecholate is assumed to be the relevant frontier orbital. The natural logarithm of the k_{cat} is proportional to the activation energy of the rate-limiting step and is correlated to the energy of HOMO [50].

The main interaction of catechol 1,2-dioxygenase from *Acinetobacter calcoaceticus* ADP1 with the substrate involves residues Leu73, Pro76, Ile105, Pro108, Leu109, Arg221, Phe253 and Ala254. Vetting and Ohlendorf [47] postulate that Arg221 may play an important role in proper positioning of the substrate's hydroxyl groups to the iron by van der Waals interactions with the aromatic ring. These interactions prevent ring rotation into two equatorial positions [47]. Matera et al. [50] observed that Arg217, Gln236 which play role in the deprotonation of substrate, together with Gly104, Pro105 and Tyr106 are required for the proper placing of aromatic compounds into the active site of catechol 1,2-dioxygenase from Gram- positive strain 1CP. Moreover residue Tyr106 forms a hydrogen bond with Tyr196 during its dissociation from the iron after the binding of catechol.

Catechol derivatives with the substituents on the C3 position can be bound in two different orientations in the enzyme active site from ADP1 strain. The substituent may be located in a pocket created by residues Ile105, Pro108, His226, Gln240 and Arg221 or its positioning may lead to a steric conflict with Tyr200. However, rearrangement in the active site makes possible the interaction of substituent on C3 with the π electrons of Tyr200 [47]. According to Matera et al. [50] the lower affinity of catechol 1,2-dioxygenase from 1CP strain for 3-substituents results from the lack of Tyr106. Tyr 196 is not connected then and it can cause the more open conformation of the space to the 3-substituent binding. It influences a certain degree of disorder in substrates binding to the iron and/or a not suitable coordination for productive conversion.

Ferraroni et al. [66] showed that preference of chlorocatechol dioxygenases for 3-chlorocatechol is connected with Val53 and Tyr78 residues whereas 4-chlorocatechols are cleaved by dioxygenases with Leu and Ile at 49 and 74 position respectively. Moreover, cysteine residues are important for interacting with chlorine substituents.

Presence of large substituents makes impossible active binding of the aromatic substrates to the active site because they interact unfavorably with Pro76, Leu73 and Ile105. Vetting and Ohlendorf [47] suggest that catechol derivatives are ligated the iron through only one of the hydroxyl group or they are trapped in an initial binding state.

3.2. Hydroxyquinol 1,2-dioxygenases characterization

Hydroxyquinol 1,2-dioxygenases have very similar structure and reaction mechanism to catechol 1,2-dioxygenases. They form homodimers with molecular weight of 58-85 kDa. The iron (III) is coordinated by two molecules of tyrosine, two molecules of histidine and one molecule of water [17,41,84-86,90].

However, Ferraroni et al. [17] revealed that His42 from H2 helix of the linker domain is connected with the copper ion in an oxidation state of +1. Probably it plays a role in stabilization of the enzyme quaternary structure. Moreover, a few different amino acid residues of hydroxyquinol 1,2-dioxygenases responsible for substrate selection were selected: Leu80, Asp83, Val107, Phe108, Gly109, Pro110, Phe111, Ile199, Pro200, Arg218, and Val251 [17].

One of the most characteristic feature of hydroxyquinol 1,2-dioxygenases is the extensive opening of the upper part of catalytic cavity which is responsible for favorable binding of hydroxyquinol [17,83]. Zaborina et al. [83] connected this fact with lower activation energy. Catalytic process of *ortho* cleavage proceeds through semiquinone radical. This intermediate can be stabilized by mesomeric delocalization of the unpaired electron in the 2-, 4-, or 6-position of semiquinone ring. With a hydroxyquinol as a substrate additional mesomeric delocalization of the unpaired electron is possible in position 1. The increase in the semiquinone character leads to a reduction of activation energy [83].

3.3. Protocatechuate 3,4-dioxygenases characterization

Protocatechuate 3,4-dioxygenases catalyze cleavage of protocatechuic acid and its derivatives. The mechanism of this reaction is the same as intradiol fission of catechol ring [13,92]. During substrate binding as a dianion to the active site of the enzyme the axial tyrosine is dissociated from the iron(III). It causes conformational changes that yields the chelated iron(III)-protocatechuate complex. Vetting et al. [16] suggested that the function of the structural rearrangements would be to facilitate the reaction of oxygen with the substrate. Electron transfer from the substrate to dioxygen aids a binding of dioxygen to the iron(III). The activation of substrate for an electrophilic attack by dioxygen leads to the formation of a peroxo bridge between the iron and C4 of substrate. O-O bond cleavage occurs after acyl migration to the peroxo oxygen. It leads to the cyclic anhydride formation and its hydrolysis to the final acyclic product [16,80,92,93].

Although protocatechuate 3,4-dioxygenases are heterodimers with molecular mass of 97.0-700.0 kDa [12,18,72,75,82] theirs tertiary structure is similar to the enzymes from catechol 1,2-dioxygenase family. Protocatechuate 3,4-dioxygenases belong to the non-heme iron dioxygenases and are composed of equimolar amounts of two different α and β subunits [11, 68,78,82,93]. These enzymes have a tetrahedral shape with α subunit forming the apex of the tetrahedron. The central cavity is accessible to solvent through triangular channels. A number of basic residues (mainly arginine and lysine) line the entrance to the active site of the enzyme producing regions of positive electrostatic potential. It is possible that these residues funnel the negatively charged substrate into the active site [16].

α and β subunits have similar folds. The core of each subunit contains two four-strand β-sheets that form a β sandwich. One β sheet consists of antiparallel strands, whereas the other has a mixed topology. The core structure is surrounded by series of small helices and large irregular loops. Two β sheets form a part of the interface between the subunits, with surrounding loops providing the remaining contacts. The amino terminus of the α subunit passes completely through a gap between the core β sandwich, supports structure of the β subunit and makes up one wall of the active-site cavity [16].

Aligments of the structure and sequences of α- and β-chain of protocatechuate 3,4-dioxygenases from more than 26 bacterial strains revealed that the β-chain show higher sequence identities than the α-chain. It may be connected with the fact that β-chain provides most of the active site [19]. However, amino acids from α-subunit may interact with substrate and in this way stabilize it in the active site. Harnett et al. [69] observed that Arg-133 from α-subunit participates in forming ionic bond with carboxyl group of protocatechuic acid.

Protocatechuate 3,4-dioxygenases from different bacterial strains are consist of variable number of αβ protomers (2 – 12) and each protomer has a cylindrical shape [12,16,78]. The minimal catalytic unit of all protocatechuate 3,4-dioxygenases is an $\alpha\beta Fe^{3+}$ structure, but $(\alpha\beta)_2Fe^{3+}$ structure has been also reported [11,72]. The crystal structure of these enzymes shows that the high-spin iron(III) is bound to the active site in a distorted trigonal bipyramidal coordination geometry with two inequivalent tyrosine ligands, two histidines, and exogenous hydroxide ion [68,69,73,77,79,92,93]. Kurahashi et al. [93] proposed that the electronic effect of the water ligand in the active site of protocatechuate 3,4-dioxygenase controls not only the distorted trigonal-bipyramidal structure but also plays a role in the oxidation of substrate. The two tyrosine ligands are thought to stabilize the ferric ion and give protocatechuate 3,4-dioxygenase its characteristic burgundy-red color via tyrosinate ligand–to-metal charge transfer [13,79].

Orville et al. [94] demonstrated that at least two iron binding sites can be occupied by exogenous ligand such as substrates, analog of substrate, and small molecules. It leads to the dynamic changes in the iron coordination sphere or the orientation of the aromatic substrate and influences the catalytic cycle. An external ligand with a higher position in the spectrochemical series ($Cl \ll RO < H_2O$) induces a larger degree of distortion [80,93].

The protocatechuate 3,4-dioxygenases show a high similarity not only for the ligands of the active site but also for the whole pocket in which the iron complex is located [77].

Orville et al. [94] studied cleavage mechanism in the presence of inhibitors. They proposed three different orientations of inhibitors in the active site. First, in which inhibitor protrudes out of the active site cavity but still coordinates the iron to yield trigonal bipyramidal coordination geometry. Second, in which inhibitor binds within the active site that is sufficient to dissociate the active site solvent molecules but without direct coordination to the iron. Third, in which inhibitor is completely bound within the active site, directly coordinates the iron, and yields octahedral geometry.

Many protocatechuate dioxygenases have narrow substrate specificities and regioselectivities [9,10,12]. Hou et al. [10] suggested that free sulfhydryl groups located at or near the active center are responsible for substrate specificity of the enzyme. It was showed that protocatechuic acid derivatives are cleaved much slower than protocatechuic acid. Transient kinetic results suggest that slower fission is probably connected with the reduced rates for substrate association and product release rather than the ring opening reaction [13].

4. Application of intradiol dioxygenases in bioremediation

Bioremediation is an effective treatment process that use organisms through their enzymatic activities. However, slow growth of microorganisms or difficulties in the control and maintain the optimal conditions for the microbial growth are disadvantages of biodegradation processes. The direct application of enzymes in the environmental treatment processes has been quite limited due to the loss of enzyme activity [95] and therefore, novel methods of enzyme stabilization are developed. One of them is immobilization which has been used as a tool to improve many of enzyme properties such as operational stability, inhibitor resistance and performance in organic solvents [95,96]. Kalogeris et al. [96] showed that immobilization of catechol 1,2-dioxygenase improved its performance at higher temperatures. It extended catechol 1,2-dioxygenase storage stability and retained its activity after operation cycles - properties which are fundamental for continuous bioprocesses during removal of toxic pollutants. Catechol 1,2-dioxygenase from *Acinetobacter radioresistens* S13 after immobilization on nanosponges showed the ability to convert the substrate, after 50 reaction cycles. Moreover, this enzyme was more thermostable [60].

Immobilized intradiol dioxygenases can also be used as biosensors because of increased interest in the detection of aromatic compounds. Zucolotto et al. [97] successfully fabricated biosensor for catechol detection. Chlorocatechol 1,2-dioxygenase was immobilized in nanostructured films in conjuction with poly(amidoamine) dendrimer in a layer-by-layer fashion. Chlorocatechol 1,2-dioxygenase remained active on these films for longer than three weeks. Moreover, this biosensor was able to detect catechol at concentration even 10^{-10} M [97].

Polychlorinated phenols, amino- and nitrophenols have become widespread environmental pollutants because of their resistance to microbial degradation and their broad toxicity. During their degradation hydroxyquinol as a key intermediate is frequently formed. Therefore, it is believed that hydroxyquinol-metabolized enzymes play an important role in the microbial degradation of various aromatic xenobiotics [51,85,88]. Immobilization of hydroxyquinol 1,2-dioxygenase from *Arthrobacter chlorophenolicus* A6 onto single-walled carbon nanotubes increased its resistance capacity to the variable environmental factors [95].

The important properties of degradation enzymes is a resistance to inhibitors such as metal ions, alcohols, chelators, hydrocarbons, phenols and other [58,59,61]. Metal ions can lead to the conformational changes in the enzymes such as a reduction in α-helices and β sheets, which can result in the loss of enzymatic activity [98] or replacement of the original metal ion at the active site of the enzymes with other metal. It may modulate their activities in accordance with the Irving-Williams studies on the stability of the metal complexes [99]. Moreover, the thiol groups of enzyme structure may bind transition metals that can deactivate the enzyme.

Gou et al. [67] observed that activity of catechol 1,2-dioxygenase from *Sphingomonas xenophaga* QYY was increased by Pb^{2+}, Mg^{2+}, K^{+}, Fe^{3+} and Ca^{2+}, whereas the addition of Mn^{2+}, Hg^{2+}, Ni^{2+} and Co^{2+} slow down the enzyme reaction. Only Cu^{2+}, Zn^{2+} and Fe^{2+} ions strongly inhibited this enzyme. Catechol 1,2-dioxygenase from *Pseudomonas aeruginosa* was not sensitive to the presence of Mg^{2+}, Fe^{2+}, and Ca^{2+}. It was inhibited by Mn^{2+}, Cu^{2+}, and Ag^{+} [64]. Ag^{+} and Hg^{2+} showed the highest inhibitory effect on catechol 1,2-dioxygenase isozymes from *Arthrobacter* sp. BA-5-17. Catechol 1,2-dioxygenase from *Rhodococcus* sp. NCIM 2891 was inhibited completely in the presence of Fe^{3+}, Cu^{2+} and Hg^{2+} [20,62]. In contrary, catechol 1,2-dioxygenase from *Gordonia polyisoprenivorans* was resistant to these cations [21]. Decreased sensitivity of both catechol 1,2-dioxygenase and protocatechuate

3,4-dioxygenase from *Stenotrophomonas maltophilia* KB2 to metal ions was observed by Guzik et al. [68]. The presence of Mn^{2+}, Ni^{2+}, Cd^{2+} caused only 20-40% reduction of protocatechuate 3,4-dioxygenase activity, whereas, catechol 1,2-dioxygenase was even activated by Cd^{2+}, Co^{2+}, Zn^{2+}, Cu^{2+}, and Fe^{2+} [68]. A high resistance of the intradiol dioxygenases to the metal ions is especially important because these ions are often found together with aromatic xenobiotics in the environment.

Some catechol 1,2-dioxgenases can be inhibited by chelating agent, especially by tiron – ferric ion chelating agent [58]. However, catechol 1,2-dioxygenase from *Arthrobacter* sp. BA-5-17 is not sensitive to chelating and sulfhydryl agents [62].

The wide substrate specificity and inhibitor resistance of some intradiol dioxygenases makes them an attractive target in the design of enzyme systems for bioremediation.

5. Conclusion

Intradiol dioxygenases are responsible for aromatic ring cleavage and for that reason are directly involved in biogeochemical cycles, and can be very useful in the development of bioremediation technology. These enzymes are the members of the aromatic-ring-cleavage dioxygenase superfamily. Based on their substrate specificity intradiol dioxygenases could be divided into three classes: catechol 1,2-dioxygenases, protocatechuate 3,4-dioxygenases, and hydroxyquinol 1,2-dioxygenase. The mechanism of intradiol fisson is common for enzymes from all clasess. In this mechanism the high-spin iron(III) is ligated by two tyrosines, two histidines, and a water ligand in trigonal bipyramidal geometry. The catalytic cycle of the intradiol dioxygenases involve binding of the catechol as a dianion and dioxygen to the metal. It leads to the formation of a peroxo bridge between iron and C4 of substrate. In the next step the Criegee rearrangement and O-O bond cleavage occurs, leading to the cyclic anhydride formation. Hydrolysis of the anhydride leads to the formation of the final acyclic product. Structural analysis of many of intradiol dioxygenases revealed only small differences in their active site structure. However, these differences significantly influence the substrate specificity and the resistance of the enzymes to the changing environmental factors.

Author details

Urszula Guzik*, Katarzyna Hupert-Kocurek and Danuta Wojcieszyńska

*Address all correspondence to: urszula.guzik@us.edu.pl

University of Silesia in Katowice, Faculty of Biology and Environmental Protection, Department of Biochemistry, Katowice, Poland

References

[1] Malawska M, Ekonomiuk A, Wiłkomirski B. Polycyclic aromatic hydrocarbons in peat cores from southern Poland: distribution in stratigraphic profiles as an indicator of PAH sources. Mires and Peat 2006;1 1-14.

[2] Li J, Yuan H, Yang J. Bacteria and lignin degradation. Frontiers of Biology in China 2009;4(1) 29-38.

[3] Greń I, Wojcieszyńska D, Guzik U, Perkosz M, Hupert-Kocurek K. Enhanced biotransformation of mononitrophenols by *Stenotrophomonas maltophilia* KB2 in the presence of aromatic compounds of plant origin. World Journal of Microbiology and Biotechnology 2010;26 289-295.

[4] Lee B-K, Vu VT. Sources, distribution and toxicity of polycyclic aromatic hydrocarbons (PAHs) in particulate matter. In: Villanyi V. (ed.) Air Pollution. Rijeka: InTech; 2010. p.99-122. Available from http://cdn.intechopen.com/pdfs/11767/InTech-Sources_distribution_and_toxicity_of_polyaromatic_hydrocarbons_pahs_in_particulate_matter.pdf (accessed 27.11.2012).

[5] Bartels I, Knackmuss H-J, Reineke W. Suicide inactivation of catechol 2,3-dioxygenase from *Pseudomonas putida* mt-2 by 3-halocatechols. Applied and Environmental Microbiology 1984;47(3) 500-505.

[6] El-Sayed WS, Ismaeil M, El-Beih F. Isolation of 4-chlorophenol-degrading bacteria, *Bacillus subtilis* OS1 and *Alcaligenes* sp. OS2 from petroleum oil-conatminated soil and characterization of its catabolic pathway. Australian Journal of Basic and Applied Sciences 2009;3(2) 776-783.

[7] Li X, Zhang T, Min X, Liu P. Toxicity of aromatic compounds to *Tetrahymena* estimated by microcalorimetry and QSAR. Aquatic Toxicology 2010;98 322-327.

[8] Zhang S, Sun W, Xu L, Zheng X, Chu X, Tian J, Wu N, Fan Y. Identification of the *para*-nitrophenol catabolic pathway, and characterization of three enzymes involved in the hydroquinone pathway, in *pseudomonas* sp. 1-7. BMC Microbiology 2012;12 27-37.

[9] Fujisawa H, Hayaishi O. Protocatechuate 3,4-dioxygenase. Crystallization and characterization. The Journal of Biological Chemistry 1968;243(10) 2673-2681.

[10] Hou ChT, Lillard MO, Schwartz RD. Protocatechuate 3,4-dioxygenase from *Acinetobacter calcoaceticus*. Biochemistry 1976;15(3) 582-588.

[11] Sterjiades R, Pelmont J. Occurrence of two different forms of protocatechuate 3,4-dioxygenase in a *Moraxella* sp. Applied and Environmental Microbiology 1989;55(2) 340-347.

[12] Hammer A, Stolz A, Knackmuss H-J. Purification and characterization of a novel type of protocatechuate 3,4-dioxyegnase with the ability to oxidize 4-sulfocatechol. Archives of Microbiology 1996;166 92-100.

[13] Elgren TE, Orville AM, Kelly KA, Lipscomb JD, Ohlendorf DH, Que L. Crystal structure and resonance Raman studies of protocatechuate 3,4-dioxygenase complexed with 3,4-dihydroxyphenylacetate. Biochemistry 1997;36 11504-11513.

[14] Murakami S, Okuno T, Matsumura E, Takenaka S, Shinke R, Aoki K. Cloning of a gene encoding hydroxyquinol 1,2-dioxygenase that catalyzes both intradiol and extradiol ring cleavage of catechol. Bioscience, Biotechnology, and Biochemistry 1999;63(5) 859-865.

[15] Iwagami SG, Yang K, Davies J. Characterization of the protocatechuic acid catabolic gene cluster from *Streptomyces* sp. strain 2065. Applied and Environmental Microbiology 2000;66(4) 1499-1508.

[16] Vetting MW, D'Argenio DA, Ornston LN, Ohlendorf DH. Structure of *Acinetobacter* strain ADP1 protocatechuate 3,4-dioxygenase at 2.2 Å resolution: implications for the mechanism of an intradiol dioxygenase. Biochemistry 2000;39 7943-7955.

[17] Ferraroni M, Seifert J, Travkin VM, Thiel M, Kaschabek S, Scozzafava A, Golovleva L, Schlömann M, Briganti F. Crystal structure of the hydroxyquinol 1,2-dioxygenase from *Nocardioides simplex* 3E, a key enzyme involved in polychlorinated aromatics biodegradation. The Journal of Biological Chemistry 2005;280(2) 21144-21154.

[18] Bubinas A, Giedraityte G, Kalediene L. Protocatechuate 3,4-dioxygeanse from thermophilic *Geobacillus* sp. strain. Biologija 2007;18(1) 31-34.

[19] Luo S, Zhang J-J, Zhou N-Y. Molecular cloning and biochemical characterization of protocatechuate 3,4-dioxygenase in *Burkholderia* sp. NCIMB 10467. Microbiology 2008;35(5) 712-719.

[20] Nadaf NH, Ghosh JS. Purification and characterization of catechol 1,2-dioxygenase from *Rhodococcus* sp. NCIM 2891. Research Journal of Environmental and Earth Science 2011;3(5) 608-613.

[21] Camargo FAO, Andreazza R, Baldoni DB, Bento FM. Enzymatic activity of catechol 1,2-dioxygenase and catechol 2,3-dioxygenase produced by *Gordonia polyisoprenivorans*. Quimica Nova 2012;35(8) 1587-1592.

[22] Li F, Zhu L. Effect of surfactant-induced cell surface modifications on electron transport system and catechol 1,2-dioxygenase activities and phenanthrene biodegradation by *Citrobacter* sp. SA01. Bioresource Technology 2012;123 42-48.

[23] Fetzner S. Ring-cleaving dioxygenases with a cupin fold. Applied and Environmental Microbiology 2012;78(8) 2505-2514.

[24] Hu X-F, Wu L. Oxidation of 3,5-di-tert-butylcatechol in the presence of V-polyoxometalate. Chemical Papers 2012;66(3) 211-215.

[25] Dorn E, Knackmuss H-J. Chemical structure and biodegradability of halogenated aromatic compounds. Biochemical Journal 1978;174 85-94.

[26] Pitter P. Correlation of microbial degradation rates with the chemical structure. Acta Hydrochimica et Hydrobiologics 1985;13(4) 453-460.

[27] Kovacs A, Hargittai I. Theoretical investigation of the additivity of structural substituent effects in benzene derivatives. Structural Chemistry 2000;11(2-3) 193-201.

[28] Pau MYM, Lipscomb JD, Solomon EI. Substrate activation for O_2 reactions by oxidized metal centers in biology. Proceedings of the National Academy of Sciences 2007;104(47) 18355-18362.

[29] Divari S, Valetti F. Caposito P, Pessione E, Cavaletto M, Griva E, Gribaudo G, Gilardi G, Giunta C. The oxygenase component of phenol hydroxylase from *Acinetobacter radioresistens* S13. European Journal of Biochemistry 2003;270(10): 2244-2253.

[30] Murray LJ, Garcia-Serres R, McCormick MS, Davydov R, Naik SG, Kim S-H, Hoffman BM, Huynh BH, Lippard SJ. Dioxygen activation at non-heme diiron centers: oxidation of proximal residue in the I100W variant of toluene/o-xylene monooxygenase hydroxylase. Biochemistry 2007;46(51): 14795-14809.

[31] Wojcieszyńska D, Greń I, Hupert-Kocurek K, Guzik U. Modulation of FAD-dependent monooxygenase activity from aromatic compounds-degrading *Stenotrophomonas maltophilia* KB2. Acta Biochimica Polonica 2011;58(3) 421-6.

[32] Greń I. Microbial transformation of xenobiotics. Chemik 2012;66 835-842.

[33] Arora PK, Srivastava A, Singh VP. Application of monooxygenase in dehalogenation, desulphurization, denitrification and hydroxylation of aromatic compounds. Journal of Bioremediation & Biodegradation 2010;1 1-112.

[34] Dresen C, Lin LY-C, D'Angelo I, Tocheva EI, Strynadka N, Eltis LD. A flavin-dependent monooxygenase from *Mycobacterium tuberculosis* involved in cholesterol catabolism. The Journal of Biological Chemistry 2010;285(29) 2264-2275.

[35] Plazmiño DE, Winkler M, Glieder A, Fraaije MW. Monooxygenase as biocatalysts: classification, mechanistic aspects and biotechnological applications. Journal of Biotechnology 2010;146(1-2) 9-24.

[36] Parales RE, Resnick SM. Aromatic ring hydroxylating dioxygenases. In: Ramos J-L, Levesque RC. (ed.) Pseudomonas. Netherlands: Springer; 2006. p287-340. Available from http://link.springer.com/chapter/10.1007/0-387-28881-3_9?null (accessed 28.11.2012).

[37] Karegoudar TB, Kim Ch-K. Microbial degradation of monohydroxybenzoic acid. The Journal of Microbiology 2000;38(2) 53-61.

[38] Guzik U, Greń I, Hupert-Kocurek K, Wojcieszyńska D. Catechol 1,2-dioxygenase from the New aromatic compounds-degrading *Pseudomonas putida* strain N6. International Biodeterioration & Biodegradation 2011;65(3) 504-512.

[39] Wojcieszyńska D, Hupert-Kocurek K, Greń I, Guzik U. High activity catechol 2,3-dioxygenase from the cresols-degrading *Stenotrophomonas maltophilia* strain KB2. International Biodeterioration & Biodegradation 2011;65(6) 853-858.

[40] Vaillancourt FH, Bolin JT, Eltis LD. The ins and outs of ring-cleaving dioxygenases. Critical Reviews in Biochemistry and Molecular Biology 2006;41(4) 241-267.

[41] Latus M, Seitz H, Eberspacher J, Lingens F. Purification and characterization of hydroxyquinol 1,2-dioxygenase from *Azotobacter* sp. strain GP1. Applied and Environmental Microbiology 1995;61(7) 2453-2460.

[42] Costas M, Mehn MP, Jensen MP, Que L. Dioxygen activation at mononuclear nonheme iron active sites: enzymes, models, and intermediates. Chemical Reviews 2004;104(2) 939-986.

[43] Mahiudddin Md, Fakhruddin ANM, Mahin AA. Degradation of phenol via meta cleavage pathway by *Pseudomonas fluorescens* PU1. International Scholarly Research Network Microbiology 2012; 2012 1-6.

[44] Mizuno S, Yoshikawa N, Seki M, Mikawa T, Imada Y. Microbial production of *cis,cis*-muconic acid from benzoil acid. Applied Microbiology and Biotechnology 1988;28(1) 20-25.

[45] Wu Ch-M, Lee T-H, Lee S-N, Lee Y-A, Wu J-Y. Microbial synthesis *cis,cis*-muconic acid by *Sphingobacterium* sp. GCG generated from effluent of a styrene monomer (SM) production plant. Enzyme and Microbial Technology 2004;35(6-7) 589-604.

[46] Gibson DT, Parales RE. Aromatic hydrocarbon dioxygenases in environmental biotechnology. Current Opinion in Biotechnology 2000;11(3) 236-243.

[47] Vetting MW, Ohlendorf DH. The 1.8 Å crystal structure of catechol 1,2-dioxygenase reveals a novel hydrophobic helical zipper as a subunit linker. Structure 2000;8(4) 429-440.

[48] Contzen M, Stolz A. Characterization of the genes form two protocatechuate 3,4-dioxygenases from the 4-sulfocatechol-degrading bacterium *Agrobacterium radiobacter* strain S2. Journal of Bacteriology 2000;182(21) 6123-6129.

[49] Pandeeti EVP, Siddavattam D. Purification and characterization of catechol 1,2-dioxygenase from *Acinetobacter* sp. DS002 and cloning, sequencing of partial *catA* gene. Indian Journal of Microbiology 2011;51(3) 312-318.

[50] Matera I, Ferraroni M, Kolomytseva M, Golovleva L, Scozzafava A, Briganti F. Catechol 1,2-dioxygenase from the Gram-positive *Rhodococcus opacus* 1CP: quantitative

structure/activity relationship and the crystal structures of native enzyme and catechols adducts. Journal of Structural Biology 2010;170(3) 548-564.

[51] Wei M, Zhang J-J, Liu H, Zhou N-Y. *para*-Nitrophenol 4-monooxygenase and hydroxyquinol 1,2-dioxygenase catalyze sequential transformation of 4-nitrocatechol in *Pseudomonas* sp. strain WBC-3. Biodegradation 2010;21(6) 915-921.

[52] Perez-Pantoja D, Donoso R, Junca H, Gonzales B, Pieper DH. Phylogenomics of aerobic bacterial degradation of aromatics. In: Timmis KN (ed.) Handbook of Hydrocarbon and Lipid Microbiology. Heidelberg: Springer Verlag; 2010. p1-40. Available from http://www.bio.puc.cl/caseb/pdf/prog7/Vol%202_00094.pdf (accessed 28.11.2012).

[53] Caposio P, Pessione E, Giuffrida G, Conti A, Landolfo S, Giunta C, Gribaudo G. Cloning and characterization of two catechol 1,2-dioxygeanse genes from *Acinetobacter radioresistens* S13. Research in Microbiology 2002;153(2) 69-74.

[54] Liu S, Ogawa N, Senda T, Hasebe A, Miyashita K. Amino acids in position 48, 52, and 73 differentiate the substrate specificities of the highly homologous chlorocatechol 1,2-dioxygenase CbnA and TcbC. Journal of Bacteriology 2005; 187(15) 5427-5436.

[55] Murakami S, Kodama N, Shinke R, Aoki K. Classification of catechol 1,2-dioxygenase family: sequence analysis of a gene for the catechol 1,2-dioxygenase showing high specificity for methylcatechols from Gram$^+$ aniline-assimilating *Rhodococcus erythropolis* An-13. Gene 1997;185(1) 49-54.

[56] Suvorova MM, Solyanikova IP, Golovleva LA. Specificity of catechol *ortho*-cleavage during *para*-toluate degradation by *Rhodococcus opacus* 1 cp. Biochemistry (Moscow) 2006;71(12) 1316-1323.

[57] Solyanikova IP, Konovalova EI, Golovleva LA. Methylcatechol 1,2-dioxygenase of *Rhodococcus opacus* 6a is a new type of the catechol-cleaving enzyme. Biochemistry (Moscow) 2009;74(9) 994-1001.

[58] Patel RN, Hou CT, Felix A, Lillard MO. Catechol 1,2-dioxygenase from *Acinetobacter calcoaceticus*: purification and properties. Journal of Bacteriology 1976;127(1) 536-544.

[59] Briganti F, Pessione E, Giunta C, Mazzoli R, Scozzafava A. Purification and catalytic properties of two catechol 1,2-dioxygenase isozymes from benzoate-grown cells of *Acinetobacter radioresistens*. Journal of Protein Chemistry 2000;19(8) 709-716.

[60] Di Nardo G, Roggero C, Campolongo S, Valetti F, Trotta F, Gilardi G. Catalytic properties of catechol 1,2-dioxygenase from *Acinetobacter radioresistens* S13 immobilized on anosponges. Dalton Transactions 2009;33 6507-6512.

[61] Sauert-Ignazi G, Gagnon J, Beguin C, Barrelle M, Markowicz Y, Pelmont J, Toussaint A. Characterization of a chromosomally encoded catechol 1,2-dioxygenase (E.C.

1.13.11.1) from *Alcaligenes eutrophus* CH34. Archives of Micorbiology 1996;166(1) 42-50.

[62] Murakami S, Wang CL, Naito A, Shinke R, Aoki K. Purification and characterization of four catechol 1,2-dioxygenase isozymes from the benzamide-assimilating bacterium *Arthrobacter* species BA-5-17. Microbiological Research 1998;153(2) 163-171.

[63] Suzuki K, Ichimura A, Ogawa N, Hasebe A, Miyashita K. Differential expression of two catechol 1,2-dioxygenases in *Burkholderia* sp. strain TH2. Journal of Bacteriology 2002;184(20) 5714-5722.

[64] Wang Ch-L, You S-L, Wang S-L. Purification and characterization of a novel catechol 1,2-dioxygenase from *Pseudomonas aeruginosa* with benzoic acid as a carbon source. Process Biochemistry 2006; 41(7) 1594-1601.

[65] König Ch, Eulberg D, Gröning J, Lakner S, Seibert V, Kaschabke SR, Schlömann M. A linear megaplasmid, p1CP, carrying the genes for chlorocatechol catabolism of *Rhodococcus opacus* 1CP. Microbiology 2000;150(Pt 9) 3075-3087.

[66] Ferraroni M, kolomytseva MP, Solyanikova IP, Scozzafava A, Golovleva LA, Briganti F. Crystal structure of 3-chlorocatechol 1,2-dioxygenase key enzyme of a new modified *ortho*-pwthway from the Gram-positive *Rhodococcus opacus* 1CP grown on 2-chlorophenol. Journal of Molecular Biology 2006;360(4) 788-799.

[67] Gou M, Qu YY, Zhou JT, Li A, Uddin MS. Characterization of catechol 1,2-dioxygenase from cell extracts of *Sphingomonas xenophaga* QYY. Science in China Series B: Chemistry 2009;52(5) 615-620.

[68] Guzik U, Hupert-Kocurek K, Sałek K, Wojcieszyńska D. Influence of metal ions on bioremediation activity of protocatechuate 3,4-dioxygenase from *Stenotrophomonas maltophilia* KB2. World Journal of Microbiology and Biotechnology 2012; DOI 10.1007/s11274-012-1178-z.

[69] Hartnett Ch, Neidle EL, Ngai K-L, Ornston N. DNA sequences of genes encoding *Acinetobacter calcoaceticus* protocatechuate 3,4-dioxyegnase: evidence indicating shuffling of genes and of DNA sequences within genes during their evolutionary divergence. Journal of Bacteriology 1990;172(2) 956-966.

[70] Kahng Y-Y, Cho K, Song S-Y, Kim S-J, Leem S-H, Kim SI. Enhanced detection and characterization of protocatechuate 3,4-dioxygenase in *Acinetobacter lwoffii* K24 by proteomics using a column separation. Biochemical and Biophysical Research Communications 2002;295(4) 903-909.

[71] Whittaker JW, Lipscomb JD, Kent TA, Munck E. *Brevibacterium fuscum* protocatechuate 3,4-dioxygenase. Purification, crystallization, and characterization. The Journal of Biological Chemistry 1984;259(7) 4466-4475.

[72] Bull Ch, Ballou DP. Purification and properties of protocatechuate 3,4-dioxygenase from *Pseudomonas putida*. A new iron to subunit stoichiometry. The Journal of Biological Chemistry 1981;256(24) 12673-12680.

[73] Walsh TA, Ballou DP. Halogenated protocatechuates as substrates for protocatechuate dioxygenase from *Pseudomonas cepacia*. The Journal of Biological Chemistry 1983;258(23) 14413-14421.

[74] Ludwig ML, Weber LD, Ballou DP. Characterization of crystals of protocatechuate 3,4-dioxygenase from *Pseudomonas cepacia*. The Journal of Biological Chemistry 1984;259(23) 14840-14842.

[75] Zylstra GJ, Olsen RH, Ballou DP. Genetic organization and sequence of the *Pseudomonas cepacia* genes for the alpha and beta subunits of protocatechuate 3,4-dioxygenase. Journal of Bacteriology 1989;171(11) 5915-5921.

[76] Stanier RY, Ingraham JL. Protocatechuic acid oxidase. The Journal of Biological Chemistry 1954;210 799-808.

[77] Petersen EI, Zuegg J, Ribbons DW, Schwab H. Molecular cloning and homology modeling of protocatechuate 3,4-dioxygenase from *Pseudomonas marginata*. Microbiological Research 1996;151(4) 359-370.

[78] Frazee RW, Livingston DM, LaPorte DC, Lipscomb JD. Cloning, sequencing, and expression of the *Pseudomonas putida* protocatechuate 3,4-dioxygenase genes. Journal of Bacteriology 1993;175(19) 6194-6202.

[79] Orville AM, Lipscomb JD, Ohlendorf DH. Crystal structures of substrate and substrate analog complexes of protocatechuate 3,4-dioxygenase: endogenous Fe^{3+} ligand displacement in response to substrate binding. Biochemistry 1997;36(33) 10052-10066.

[80] Valley MP, Brown CK, Burk DL, Vetting MW, Ohlendorf DH, Lipscomb JD. Roles of the equatorial tyrosyl iron ligand of protocatechuate 3,4-dioxygenase in catalysis. Biochemistry 2005;44(33) 11024-11039.

[81] Overhage J, Kresse AU, Priefert H, Sommer H, Krammer G, Rabenhorst J, Steinbüchel A. Molecular characterization of the genes *pcaG* and *pcaH*, encoding protocatechuate 3,4-dioxygenase, which are essential for vanillin catabolism in *Pseudomonas* sp. strain HR199. Applied and Environmental Microbiology 1999;65(3) 951-960.

[82] Chen YP, Dilworth MJ, Glenn AR. Aromatic metabolism in *Rhizobium trifolii* – protocatechuate 3,4-dioxygenase. Archives of Microbiology 1984;138(3) 187-190.

[83] Zaborina O, Seitz H-J, Sidorov I, Eberspächer J, Alexeeva E, Golovleva L, Lingens F. Inhibition analysis of hydroxyquinol-cleaving dioxygenases from the chlorophenol-degrading *Azotobacter* sp. GP1 and *Streptomyces rochei* 303. Journal of Basic Microbiology 1999;39(1) 61-73.

[84] Daubaras DL, Saido K, Chakrabarty AM. Purification of hydroxyquinol 1,2-dioxygenase and maleylacetate reductase: the lower pathway of 2,4,5-trichlorophenoxyacetic

acid metabolism by *Burkholderia cepacia* AC1100. Applied and Environmental Microbiology 1996;61(11) 4276-4279.

[85] Takenaka S, Okugawa S, Kadowaki M, Murakami S, Aoki K. The metabolic pathway of 4-aminophenol in *Burkholderia* sp. strain AK-5 differs from that of aniline and aniline with C-4 substituents.

[86] Travkin VM, Jadan AP, Briganti F, Scozzafava A, Golovleva LA. Characterization of an intradiol dioxygenase involved in the biodegradation of the chlorophenoxy herbicides 2,4-D and 2,4,5-T. FEBS Letters 1997;407(1) 69-72.

[87] Louie TM, Webster ChM, Xun L. Genetic and biochemical characterization of a 2,4,6-trichlorophenol degradation pathway in *Ralstonia eutropha* JMP134. Journal of Bacteriology 2002;184(13) 3492-3500.

[88] Hatta T, Nakano O, Imai N, Takizawa N, Kiyohara H. Cloning and sequence analysis of hydroxyquinol 1,2-dioxygenase gene In 2,4,6-trichlorophenol-degrading *Ralstonia pickettii* DTP0602 and characterization of its product. Journal of Bioscience and Bioengineering 1999;87(3) 267-272.

[89] Yoshida M, Oikawa T, Obata H, Abe K, Mihara H, Esaki N. Biochemical and genetic analysis of the γ-resorcylate (2,6-dihydrobenzoate) catabolic pathway in *Rhizobium* sp. strain MTP-10005: identification and functional analysis of its gene cluster. Journal of Bacteriology 2007;189(5) 1573-1581.

[90] Kitagawa W, Kimura N, Kamagata Y. A novel *p*-nitrophenol degradation gene luster from a gram-positive bacterium, *Rhodococcus opacus* SAO101. Journal of bacteriology 2004;186(15) 4894-4902.

[91] Mayilmurugan R, Sankaralingam M, Suresh E, Palaniandavar M. Novel square pyramidal iron(III) complexes of linear tetradentate bis(phenolate) ligands as structural and reactive model for intradiol-cleaving 3,4-PCD enzymes: quinone formation *vs.* intradiol cleavage. Dalton Transactions 2010;39 9611-9625.

[92] Borowski T, Siegbahn PEM. Mechanism for catechol ring cleavage by non-heme iron intradiol dioxygenases: a hybrid DFT study. Journal of American Chemical Society 2006;128(39) 12941-12953.

[93] Kurahashi T, Oda K, Sugimoto M, Ogura T, Fujii H. Trigonal-bipyramidal geometry induced by an external water ligand In a sterically hindered iron salen complex, related to the active site of protocatechuate 3,4-dioxygenase. Inorganic Chemistry 2006;45(19) 7709-7721.

[94] Orville AM, Elango N, Lipscomb JD, Ohlendorf DH. Structures of competitive inhibitor complexes of protocatechuate 3,4-dioxygenases: multiple exogenous ligand binding orientations within the active site. Biochemistry 1997;36(33) 10039-10051.

[95] Suma Y, Kim D, Lee JW, Park KY, Kim HS. Degradation of catechol by immobilized hydroxyquinol 1,2-dioxygenase (1,2-HQD) onto single-walled carbon nanotubes. In-

ternational Conference on Chemical, Environmental Science and Engineering (ICE-EBS'2012), July 28-29, 2012, Pattaya, Thailand.

[96] Kalogeris E, Sanakis Y, Mamma D, Christakopoulos P, Kekos D, Stamatis H. Properties of catechol 1,2-dioxygenase from *Pseudomonas putida* immobilized in calcium alginate hydrogels. Enzyme and Microbial Technology 2006;39(5) 1113-1121.

[97] Zucolotto V, Pinto APA, Tumolo T, Moraes ML, Baptista MS, Riul A, Araujo APU, Oliveira ON. Catechol biosensing using a nanostructured layer-by-layer film containing Cl-catechol 1,2-dioxygenase. Biosensors and Bioelectronics 2006;21(7) 1320-1326.

[98] Latha R, Mandappa IM, Thakur MS, Manonmani HK. Influence of metal ions on dehydrohalogenase activity. African Journal of Basic & Applied Sciences 2011;3(2) 45-51.

[99] Gopal B, Madan LL, Betz SF, Kossiakoff AA. The crystal structure of a quercetin 2,3-dioxygenase from *Bacillus subtilis* suggests modulation of enzyme activity by a change in the metal ion at the active site(s). Biochemistry 2005;44(1) 193-201.

Chapter 8

Hydrocarbon Biodegradation Potential of Native and Exogenous Microbial Inocula in Mexican Tropical Soils

Ildefonso Díaz-Ramírez, Erika Escalante-Espinosa,
Randy Adams Schroeder ,
Reyna Fócil-Monterrubio and Hugo Ramírez-Saad

Additional information is available at the end of the chapter

http://dx.doi.org/10.5772/56233

1. Introduction

Soil and water contamination by oil is one of the central environmental problems in Mexico. In the southeastern part of the country, especially in Tabasco and Veracruz states there are several oil facilities involved in extraction, transportation, processing and storage of oil and oil products. Historically in this region there have been spills that affect large areas of soil and eventually rivers, streams and lagoons. Pollutants often persist for long periods in the soil, sediments and water due to different factors like the environmental conditions prevailing in the region (heavy rains and extensive wetlands), the nature of the soils (mostly clayed soils adjacent to water bodies), and the recalcitrance of oil components (Adams et al. 2011).

The application of bioremediation methods based on degradation of hydrocarbons is a widely used alternative for the recovery of such contaminated soils (Boonchan et al. 2000, Sayara et al. 2012). However, the success of this strategy depends on the degrading capacity of the native or exogenous microorganisms applied as part of the treatment (Liu and Suflita, 1993, Van Hamme, 2003, Liu et al. 2011). In this sense, a key issue for the restoration of soils is to assess their potential for bioremediation under different treatment strategies and culture conditions. This analysis must consider first the biodegrading capacity of native microbial populations (natural attenuation). Alternatively, the use of inoculants composed by exogenous microorganisms, previously characterized as degraders (bioaugmentation), can be considered as an active bioremediation treatment.

Several authors have reported that specific microorganisms are required for the removal of various petroleum fractions, in particular those having enzymatic capabilities for biotransformation of these compounds (Van der Meer et al. 1992; Prince, 2003). Although the bacteria can be grown in the laboratory, progress on knowledge of microbial ecology of complex communities requires more extensive studies, regarding the activity of microorganisms in environmental samples that suffered minimal modifications (Amann and Kühl, 1998). Moreover, knowledge of the mechanisms of degradation of contaminants in the environment has been generated in laboratory studies using degraders. As a result the role of microorganisms in regulating processes such as mineralization and biodegradation of polluting compounds is not fully understood (Watanabe and Hamamura, 2003).

1.1. Adaptation and presence of microorganisms in contaminated sites

Exposure of native microbial soil populations to hydrocarbon contamination is a key factor that determines the biodegradation rates for overall removal of contaminant. This phenomenon allows the increase in oxidation potential of the microbial community and is called adaptation. According to Leahy and Colwell (1990), this term applies to mixed microbial communities and single microbial species. Van der Meer et al. (1992) proposed some mechanisms on the molecular events that allow adaptation of microbial communities: i) induction and/or de-repression of specific enzymes, ii) new metabolic capabilities as a result of genetic changes, and iii) enrichment of microbial populations capable of transforming the compound of interest.

Generally, degrading-microorganisms are more frequently isolated from hydrocarbon contaminated soils, where they can reach levels between 1 to 10% of the soil microbial population. While in non polluted environments degraders normally represent 1% of the microbial community (Torsvik and Ovreas, 2002). In some cases, predominant genera of degraders are reported, while in others there is a high microbial diversity, which may be due to the composition of the native microbial community, and the environmental conditions at the affected sites. Microbial ability to grow in the presence of oil-derived compounds as a sole carbon source has been addressed by selective microbial enrichment studies (Leahy and Colwell, 1990, Viñas et al. 2002, Roldán-Carrillo et al. 2012).

1.2. Hydrocarbons biodegradation

The biodegradation process is defined as the change in the chemical structure of a compound carried out by microorganisms, and is the basis of conventional engineering techniques for the treatment of sewage and contaminated soil bioremediation (Madsen, 1998). The petroleum compounds can selectively inhibit certain microorganisms, causing changes in the number and diversity of microbial species in soil. Thus the rate of biodegradation of hydrocarbon is determined by the presence of such native microorganisms, their physiological capabilities and the environmental conditions (Prince, 2003, Zhang et al. 2012).

1.2.1. Applying degraders

In some cases, hydrocarbon biodegradation levels of native inocula had been higher than commercial or exogenous inocula (Thouand et al. 1999, Mohammed et al. 2007). It is considered that exogenous inocula must possess two essential properties: i) to degrade oil faster than native populations, and ii) to degrade a broader spectrum of oil compounds than native populations (Van Hamme et al. 2003, Madueño et al. 2011). Obtaining microorganisms that meet these conditions may be more feasible from sites where the microbial population is exposed for long periods to hydrocarbons pollutants (Liu and Suflita, 1993, Greenwood et al. 2009).

1.2.2. Biodegradation of hydrocarbons by mixed cultures

It is generally considered that mixed cultures have a higher metabolic versatility than pure cultures (axenic). Mixed cultures have been used as inocula for treating waste oil in various systems, applying preadaptation strategies of consortia to different types of hydrocarbons and field application conditions (Van Hamme et al. 2000). Some reports have studied the biodegrading abilities of defined mixed cultures where the composition of the consortia is previously known. Ko and Lebeault (1999), reported the co-oxidation of decalin and pristane in the presence of hexadecane, by a defined mixed culture of two bacterial strains. Finding that decalin was degraded to a greater extent by an axenic strain than the mixed culture, while pristane was largely degraded by the mixed culture. Richard and Vogel, (1999) compared the biodegradation of diesel by three bacterial strains and a mixed culture composed of the same, degradation of aliphatics was greater with the mixed culture than with the single strains, these authors suggest that degradation of hydrocarbon mixtures by mixed cultures may involve different strains that possess complementary metabolic capabilities.

Often the success of the addition of exogenous microorganisms depends on their adaptation to the environmental conditions prevailing in the soil and their ability to compete with native microorganisms. However some cases have been documented where native populations are better degraders than exogenous inoculants (Thouand et al. 1999). An alternative is the application of microorganisms native of sites contaminated with the compounds of interest, they can pre-selected based on their metabolic capabilities and subsequently be characterized in the laboratory for possible use (Medina-Moreno et al. 2005; Diaz -Ramirez et al. 2008). The application of such microorganisms is useful when the number of native microorganisms is low and they do not have a proper biodegrading capacity (Van Hamme et al. 2003). In general, bacterial cocktails used as inoculants are undefined mixed cultures, where the composition and metabolic capacities of the microbial members is not known precisely; therefore the role of each strain during biodegradation is not clear.

Another useful alternative for the treatment of oil contaminated soils is the use of native plants together with degrading microorganisms associated in their rhizosphere, known as phytoremediation or phytomitigation strategies. The use of plants for remediation of contaminated soil is well accepted, particularly when they are part of a sequence of treatments. The use of pastures allows overall improvement of soil fertility conditions and biological activity (enzymatic functions and respiratory activity), to reduce toxicity, improve, mineralization and / or integration of polluting compounds via humification (Escalante-Espinosa, 2005; Adams et al. 2011).

Diaz-Ramirez et al. (2003) reported a strategy involving bacterial strains obtained from the rhizosphere of a plant native to a contaminated site, that were used to prepare defined mixed cultures for hydrocarbon biodegradation and mineralization in liquid cultures. Ten bacterial strains were isolated and identified as members of the genera: *Bacillus, Gordonia, Kocuria, Pseudomonas, Arthrobacter* and *Micrococcus.* Then assessed for biodegradation of specific hydrocarbon fractions in liquid medium, as both; single strains and defined mixed cultures (CMD) prepared with standardized mixtures. Additionally predominant strains were selected and their population dynamics during biodegradation was analyzed, in order to design specific inocula for different types of hydrocarbons (Diaz-Ramirez et al. 2008). Under this approach consumption levels of 40% of total hydrocarbons were reached in 11 days of culture, degradation value for aromatics was 31%, and 20% for polar hydrocarbons. The results pointed that co-oxidation of aromatic and polar compounds was possible in the presence of aliphatic fraction, when using a defined mixed culture consisting of 3 *Bacillus* strains, 2 *Pseudomonas* strains and one *Gordonia* isolate (Diaz-Ramirez et al. 2008).

1.3. Monitored natural attenuation

Conventional bioremediation techniques generally allow the recovery of soils affected by organic pollutants. Natural attenuation has been considered an important remediation strategy mainly when soil, groundwater or surface water is involved, in places where biogeochemical conditions favor natural processes that degrade or immobilize harmful contaminants. Natural attenuation includes processes such as biodegradation, dispersion, sorption and volatilization of contaminants (Boettcher and Nyer, 2001). Monitored Natural Attenuation (MNA) is a viable alternative for reduction of toxicity, mobility or concentration of the contaminant compounds in soil and/or water (EPA, 1999). Some authors have noted that natural attenuation processes can be selected alone or in conjunction with more active alternatives, when achieving the objectives of remediation, in reasonable time periods, compared with other alternatives that can be costly or with high impact to ecosystems. Among the parameters that should be determined in order to assess the biodegradation potential of the microbial community are: the number and type of predominant microorganisms, degradation kinetics (measured in field or lab), enzyme activities (dehydrogenases, lipases, oxygenases), as well as the half-life of contaminants, calculated from experimental data based on zero- or first order kinetics (Boettcher and Nyer 2001, Margesin et al. 2005, Chang et al. 2010).

In this chapter, we present two experimental approaches developed to assess the ability of native and exogenous microbial populations to biodegrade different types of hydrocarbons, under lab and field conditions.

In the first experimental approach, the degrading activity of native and exogenous microorganisms on diesel and Olmeca crude oil was assessed, under different culture conditions at laboratory scale. All exogenous microorganisms were previously isolated from the rhizosphere of *Cyperus laxus* Lam., a plant native of a highly contaminated tropical swamp (Diaz-Ramirez et al., 2003, Escalante-Espinosa, et al. 2005). The biodegradation ability of these microorganisms was tested in soil artificially contaminated with diesel and biostimulated.

http://dx.doi.org/10.5772/56233

Further biodegradation experiments in soil contaminated with crude Olmeca oil, were carried out in the presence and absence of co-culture.

The second experimental approach involved the monitoring of natural attenuation (MNA) as a result of a gasoil spill occurred in an area of ecological value; a riparian environment consisting of two streams and two major rivers located in the southeast of Mexico. We identified five points affected by gasoil on areas where the stream presented meanders and points of plant material accumulation. The initial TPH concentrations ranged from 8 500 to 60 000 mg kg^{-1} soil at different points, total hydrocarbon content was determined periodically in order to follow the natural attenuation process.

2. Material and methods

Experiments were carried out with tropical soils, artificially contaminated with either diesel or Olmeca crude oil. In both experimental designs total hydrocarbon content was determined at different sampling times (biodegradation kinetics), the overall biodegradation rate, the total number of microorganisms and maximum number of degraders were calculated. The composition of the residual diesel was assessed by gas chromatography and for Olmeca crude oil by column chromatography. The half-life of hydrocarbons in soil and the biodegradation constant using a first order model were calculated for both biodegradation assays.

In addition, another study was conducted in order to assess the natural attenuation potential of soil and sediments of a stream system affected by a gasoil spill. Periodic sampling was carried out and the analysis of kinetic parameters suggested alternatives that would allow restoration of different affected areas.

2.1. Soil samples

Soil was collected close but outside of an area highly impacted by human activities (truck parking lot), with high deposition of plant material and therefore high microbial activity potential. The soil samples showed dark color with fine texture (silt-clay) and pH 7. Soil humidity at the time of collection was 30% with a water retention capacity of 36% (w / w). Proportions of clay, silt and sand were 40%, 45% and 15%, respectively. Soil organic matter content was 2.88%, corresponding to a soil with medium organic matter content (NOM-021-SEMARNAT-2000); these characteristics are shared by many soils of the study area, considered representative of tropical soils of southeastern Mexico.

2.2. Diesel biodegradation

Biodegradation experiments of diesel (10,000 ppm) were performed in shaken flasks with liquid medium (adjusting the ratio C/ N ~ 20). In such assays, we tested the degrading activity of two defined mixed cultures, the first one (B-DMC) consisted of six bacterial strains; *Bacillus cereus, Gordonia rubripertincta, Kocuria rosea, Bacillus subtilis* strains 7A and 9A, and a non-identified strain UAM10AP. The second defined mixed culture was formed by two fungal strains (F-DMC); *Aspergillus terreus* and *Aspergillus carneus*. Finally, a co-culture (CC) composed of all of the above strains was also tested. The degrading microorganisms were isolated and characterized as previously described (Diaz-Ramirez et al, 2003). In parallel, non-inoculated controls were prepared to quantify the hydrocarbons abiotic loss.

In order to evaluate the biodegradability of diesel by exogenous and native hydrocarbon-degrading microorganisms, an experimental trial was performed using artificially contaminated soil (~ 120,000 ppm of diesel), which has been biostimulated by addition of a commercial fertilizer (Triple 17), in order to get an initial C/N ratio of 20. Two treatments were established for this assay; a) bioaugmented soil (inoculated with exogenous microorganisms), and b) non-bioaugmented soil (only with the native microbiota).

2.2.1. Gas chromatography analysis

Analytical procedure to determine residual hydrocarbons was based on the gas chromatography standard method EPA8015B. A gas chromatograph (Agilent Technologies Mod 6850 – II) equipped with FID and a column Agilent HP-1. Running conditions were Initial Temp: 70°C, Initial Hold 4.00 min, Ramp 1: 10°C/min to 70°C. Ramp 2: 40°C/min to 310°C, hold for 9.00 min. Injector temp: 250°C. Detector temp: 300°C. Helium was the carrying gas at: 10.00 psi. For calibration, standards were prepared with hydrocarbons known to be present in the used diesel, obtaining a standard curve with 8 quantification points.

2.3. Olmeca crude oil biodegradation

Assays were performed in trays containing 400 g of artificially contaminated soil (10,000 mg TPH kg^{-1} soil), amended (biostimulated) with 2% w/w sugar cane bagasse used as organic conditioner, and the commercial fertilizer Gro-green 20:30:10 (Campbell, Mexico) to obtain a C / N initial ratio of 6. Soil humidity was maintained at 40% of the water retention capacity of soil. Four treatments were established as follow: a) Bioaugmented - sterile soil (B-SS), b) Non bioaugmented - sterile soil (NB-SS), c) Bioaugmented - non-sterile soil (B-NSS), and d) Non bioaugmented - non-sterile soil (NB-NSS).

The following variables were determined: TPH content at different sampling times (EPA 3540 method), the number of hydrocarbon-degrading bacteria was assessed by the most probable number assay (Wrenn and Venosa, 1996), while total heterothophic microorganisms were determined after Lorch et al (1995). With the data regarding residual TPH content, the biodegradation extent was calculated, as well as the global and maximum consumption rates. Additionally, kinetic parameters were calculated as the adjustment of the consumption data to the first order model (as described above in the section 3.2). Biodegradation kinetics of diesel in soil), the biodegradation constants and half-life under different treatments were also obtained.

2.4. Monitoring of natural attenuation

A periodic sampling protocol was designed to monitoring the natural attenuation of a soil highly contaminated with gasoil (medium fraction petroleum). Based on criteria as site geomorphology and direction of the stream meanders; several points with presence of hydrocarbons were identified and selected for monitoring. The residual gasoil content (medium fraction C10-C28) was used as a measure of passive biodegradation or Monitored Natural Attenuation.

The residual gasoil content (medium fraction C10-C28) was used as a measure of passive biodegradation or Monitored Natural Attenuation. Soil hydrocarbon analyses were performed by Gas Chromatography accordingly to the Mexican regulation (NOM-138-SEMARNAT/SS-2003), which is based on the EPA 8015b standard procedure. From these values, biodegradation extent and half life of the contaminant in soil were calculated. In addition, the amounts of oil-degrading and heterotrophic microorganisms were determined by

3. Results and discussion

To assess the natural attenuation potential of contaminated soils, as well as to develop inocula for bioremediation of such sites, it is necessary to know the degrading abilities of microbial native populations. Where applicable, the persistence and activity of the introduced microorganisms, without bioavailability limitations attributable to the soil, have to be addressed. Contrast the biodegradation process under this scheme and in field is one of the main challenges for the definition of the auto recovery potential of contaminated soils, as well as for the establishment of bioremediation strategies that are economic and technically feasible.

3.1. Biodegradation of diesel in liquid cultures

Higher levels of biodegradation were obtained with the CC and F -DMC that behave very similar (Table 1). Statistical analysis did not show significant differences at 95% confidence among these two treatments. While the consumption rate data also showed similar values for these treatments. Similar results were obtained with the previously reported CMD (six bacteria) using aliphatics as carbon source (10 000 mg L^{-1}) obtained from Maya crude oil, consumption rates of 282 mg L^{-1} d^{-1} in eleven days of culture were reported (Diaz-Ramirez et al. 2008).

The number of viable microorganisms was higher in cultures in liquid medium with the standardized mixture of bacterial strains and co-culture (0.36 - 8.4 ± 1.1 X10^8 UFC ml^{-1}) compared to the initial values (<1x10^7 CFU ml^{-1}).

These results indicated that the application of the standardized mixture of microorganisms (bacteria - fungal strains) was favorable in terms of overall growth. Considering the high biodegradation results, probably there was a synergistic effect between the populations present in the CC and F-DMC, as it has been previously reported with similarly defined mixed culture in liquid medium (Diaz-Ramirez et al. 2003).

Microbial inocula	Biodegradation extent (%)	Consumption rate (mg $L^{-1}d^{-1}$)
B-DMC	36.5 ± 6.8 a	156.63 ± 48.6
F-DMC	49.1 ± 2.8 b	204.88 ± 12.0
CC	48.3 ± 5.2 b	201.33 ± 22.0

Table 1. Diesel biodegradation by the different microbial inocula assayed during 16 days of incubation. Biodegradation extent was calculated considering an initial TPH concentration of 0.5 g/flask and corrected considering the diesel abiotic loss in the control. Values followed by the same letter did not show significant differences.

3.2. Biodegradation kinetics of diesel in soil

In order to determine the biodegradation level of diesel in soil and the activity of the exogenous and native microbial populations, the residual oil content was determined during the biodegradation assays, allowing to perform a kinetic study. Additionally, we quantified the number of heterotrophic microorganisms and soil toxicity at the end of treatment.

Figure 1 shows the degradation kinetics of diesel in artificially contaminated soil (120,000 mg kg^{-1} soil), determined as diesel residual hydrocarbons.

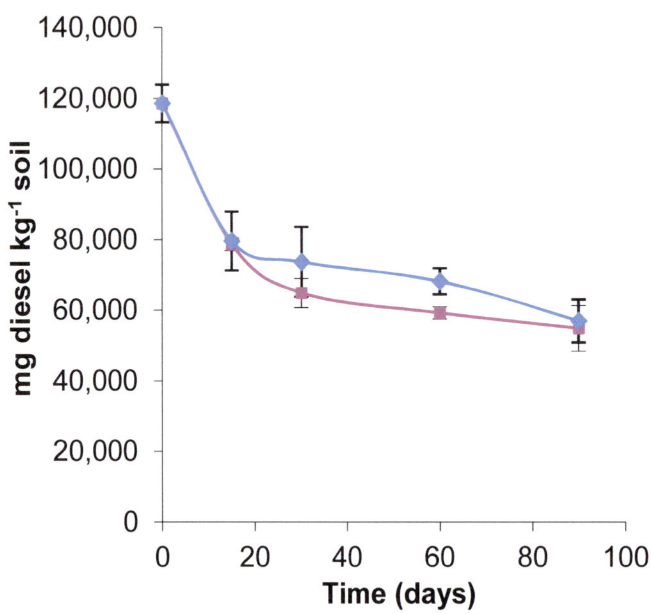

Figure 1. Soil hydrocarbon content throughout 90 day-assay, initial concentration was 120,000 mg Kg^{-1} for the bioaugmented (■) and non-bioaugmented soil (◆).

The higher degradation rates were observed during the first 30 days of culture, being 37.8% and 45.3% for the bioaugmented and non-bioaugmented soils, respectively. This rapid initial degradation is likely due to the consumption of the easily degradable compounds, for example C10 - C18 aliphatic hydrocarbons, and some low molecular weight aromatic compounds (Vallejo et al. 2005). Subsequently, between 30 and 90 days of culture, diesel biodegradation in bioaugmented and non- bioaugmented treatments was constant until the end of the culture (Figure 1), reaching ultimate biodegradation of 52% and 54% for bioaugmented and non-bioaugmented soil, respectively. No significant differences were found between the means of the two treatments for biodegradation of diesel oil (t test, P> 0.05).

Further analysis of these results showed that in the first 30 days of treatment there was a rapid decline in the concentration of hydrocarbons (diesel) in both; the bioaugmented and the non-bioaugmented soils, thereafter the biodegradation rate decreased. High activity of soil microbial populations has been frequently reported for biostimulated soils (Sabaté et al, 2004; Margesin, 2005, Chemlal et al. 2012). Based on colony-morphology observations, the microorganisms inoculated in the treated soil were prevailing at the end of the assay. The predominance of these hydrocarbon-degrading strains in liquid culture and solid media has been previously reported using molecular tools (Escalante-Espinosa, 2005; Diaz-Ramirez et al, 2008), confirming that inoculated strains have good capacity to remain in soils under the assayed conditions.

Based on a first order model degradation constant (-k) and half-life ($t_{1/2}$) were determined (Eweis et al, 1998). Table 2 presents the kinetic parameters calculated for the removal of diesel hydrocarbons from the assayed treatments.

Treatment	Biodegradation (%)	GCR[a] (mg kg^{-1} d^{-1})	MCR [b] (mg kg^{-1} d^{-1})	-k [c] (d^{-1})	Half life time (d)
Bioaugmented	52.0 ± 5.1	684.4 ± 67.6	2 598 ± 555	-0.0093	74.5
Non-bioaugmented	53.7 ± 5.4	707.7 ± 71.6	2 657 ±119	-0.0106	65.4

[a] Global Consumption Rate (GCR) was calculated considering the initial and residual hydrocarbon content after 90 days.

[b] Maximum Consumption Rate (MCR) was calculated as described in the kinetic parameter section below.

[c] Biodegradation constant was calculated as:

$$C = C_0 e^{-kt} \quad (1)$$

Where:

C= Total petroleum hydrocarbon concentration at "t" time

C_0= Initial concentration of petroleum hydrocarbon

k = Biodegradation constant represented as the slope of the $Ln\ C/C_0$ *vs time* (days).

Table 2. Kinetic parameters for the Diesel biodegradation assay in soil.

According to the analysis the estimated biodegradation constant was similar for both treatments. These values are comparable to those reported by Cardona and Iturbe (2003) for the biodegradation of diesel in an agricultural soil, under biostimulated conditions, adding alternate sources of oxygen and nitrogen to the soil. The values for the biodegradation constants were 0.044 d^{-1} – 0.011 d^{-1}, with concentrations ranging from 28 000 to 40 000 mg diesel kg^{-1} soil. The half-life time of spilled diesel in this soil was calculated, indicating that from 4.4 to 5 months were required to achieve a complete removal of hydrocarbons.

In similar biodegradation assays, Hye Hong et al (2005), reported 60% degradation of diesel in 13 days using a *Pseudomonas aeruginosa* (IU5 strain) isolated from an oil-contaminated soil. Marquez-Rocha et al (2001), reported rates of biodegradation (GCR) of diesel (118 g kg^{-1} dry soil) of about 2 130 to 2 780 mg kg^{-1} diesel dry soil d^{-1}, with a bacterial consortium previously acclimated to diesel. Diaz-Ramirez (2000) evaluated the biodegradation of hydrocarbons (medium fraction) using a bacterial consortium obtained from the rhizosphere of *Cyperus laxus* Lam., finding a 62% biodegradation within 30 days, the level of biodegradation was similar to the one achieved in this work; 37.8% after 30 days of treatment in the bioaugmented soil.

3.2.1. Analysis of residual diesel in soil after 90 days

Although the level of biodegradation of diesel in the bioaugmented and non-bioaugmented treatments was similar at the end of culture (Table 2), the composition profiles of the residual hydrocarbons were different in relation to the initial composition of diesel, as shown in the comparative chromatographic analysis (Figure 2).

The commercial diesel fuel used in the biodegradation assay is constituted by a mixture of paraffinic and olefinic hydrocarbons ranging from C10 to C28, plus 30% aromatics. The maximum total sulfur content is <0.05% (PEMEX, 2008). Under Mexican standards this type of diesel fraction is regarded as medium-fraction hydrocarbons, whose presence is limited to 1 200 ppm for agricultural soil, and 5 000 ppm in industrial soil (NOM-138-SEMARNAT/SS-2003).

Figure 2a shows the initial composition of diesel fuel used in both treatments; the most prominent peaks corresponded to the alkanes present in larger proportion in the mixture. In bioaugmented soil (Figure 2b), there was a marked reduction in the number of peaks and the area under the curve after 90 days, whereas in the bioaugmented soil (Figure 2c) an increase in the area under the curve was observed. This later change indicated that the residual hydrocarbons were composed of a more complex mixture (unresolved complex mixture), characteristic of a soil with residual compounds of higher molecular weight and chemical complexity. In the bioaugmented treatment the main biotransformation of hydrocarbons was probably due to degradation of short chain compounds (low molecular weight) and medium-sized alkanes (Figure 2b).

This chromatographic analysis is consistent with the results obtained by Hye et al. (2005), which shows a comparison of diesel before and after culturing with a *P. aeruginosa* strain IU5, suggesting that such a decrease is mainly due to the degradation of short to medium chain

http://dx.doi.org/10.5772/56233

aliphatics (C10 - C24). Moreover, Marquez-Rocha et al. (2001), reported the degradation of medium sized hydrocarbons (> C12) contained in the diesel, together with the short-chain compounds.

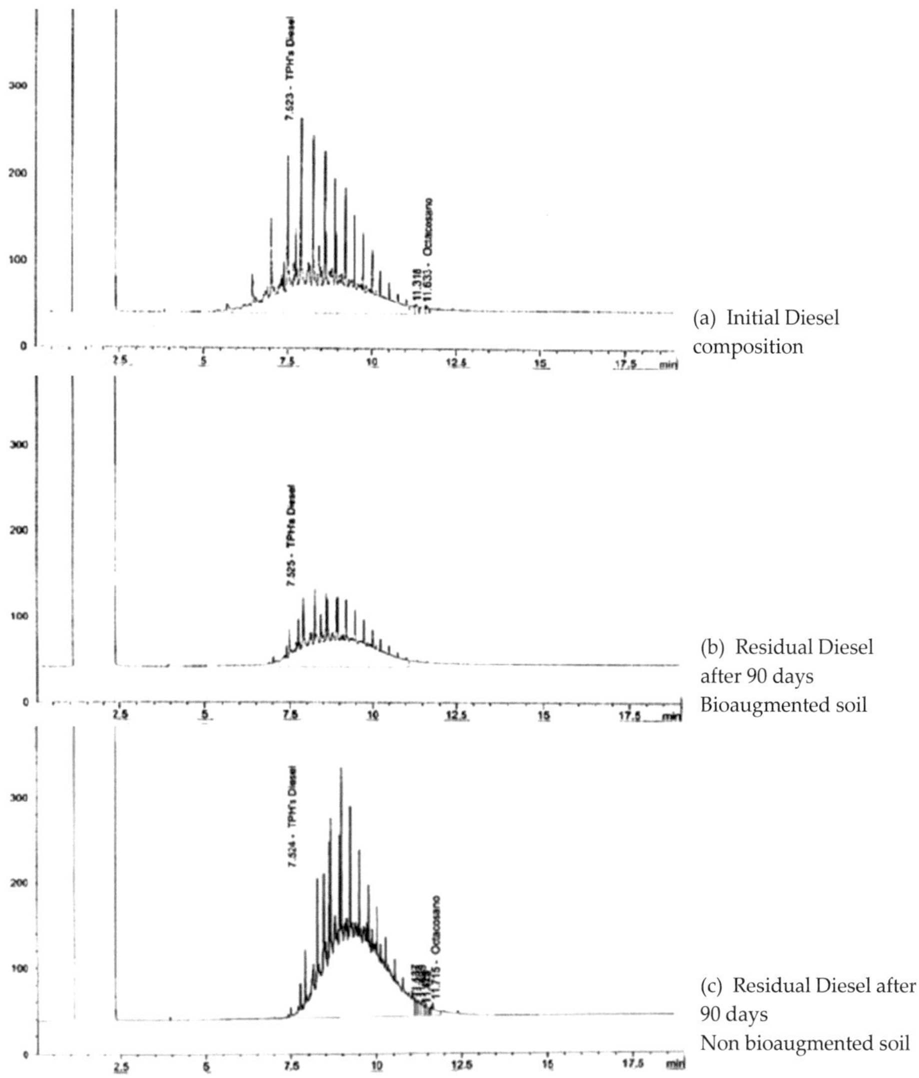

Figure 2. Gas chromatography of diesel used as carbon source after 90 days of culture on bioaugmented and non-bioaugmented artificially contaminated soil. Chromatography analysis conditions were made based on Mexican normativity (NOM-138-SEMARNAT/SS-2003) and the EPA method (EPA 8015b). Initial and final composition for both treatments is shown for comparative analysis.

3.2.2. Assessment of degrading microorganisms

The number of hydrocarbon-degrading bacteria was assessed by the most probable number assay (Wrenn and Venosa, 1996). In bioaugmented soil values were between 11.0 to 110 X 10^5 degrading microorganisms g^{-1} dry soil, and for the non bioaugmented treatment values ranged from 0.21 to 2.1 x 10^5 degraders g^{-1} dry soil. Considering that the used soil has no history of contamination, the number of diesel degraders in both treatments is high. Similar level of degraders (2.1x10^5 g^{-1} dry soil) was reported by Vallejo et al (2005) in their evaluation of biostimulation on PAH biodegradation of petroleum contaminated soils.

3.2.3. Analysis of toxicity

In order to evaluate the acute toxicity of soil due to the presence of diesel and thus the effect of microbial activity (biodegradation) during the assay, the final acute toxicity of soil samples under both treatments was addressed. The measurements were made using the Microtox ® bioassay under the Mexican regulation procedure (NMX-AA 112-SECOFI, 1995). Soil samples were prepared following the Toxicity Characteristic Leaching Procedure (TCLP) as described in NOM-053-ECOL-1993. Toxicity was calculated based on the Relative Index of Toxicity (RIT) for contaminated soils located in Mexican southeastern. RIT is a relation between the half effective concentration (EC_{50}) and the toxicity level for contaminated soils (Adams and Guzman-Osorio 2008). In tropical soils, hydrocarbons identified as medium fraction (C10 - C28) have a low toxicity in low concentrations (<10 000 ppm), mainly due to the low molecular weight compounds, with short half-life (2 - 4 months) under natural conditions. In contrast, at higher concentrations (> 10,000 ppm), toxic effects are seen together with changes in the physicochemical properties and functions of soil, like; a decrease in nutrient retention capacity and fertility, with further increases in compaction, repellency and salinization (Adams et al. 2008). Many contaminated sites in the southeast of the country have very high hydrocarbon concentrations (> 25,000 ppm) and a high degree of weathering.

The index provides a null level of toxicity for leachates of soils with more than 95 000 ppm; slightly toxic for soil leachates ranging from 84 700 to 58 900 ppm, while levels ranging between 58 900 and 36 000 ppm are regarded as a toxic. This toxicity levels can be transformed into Toxicity Units (TU= (1/EC_{50}) x10^6). The EC_{50} is the concentration causing a 50% decrease in the bioluminescence level of test organism.

Toxicity results are presented in Table 3, soil samples of both treatments showed a toxicity of about 15 TU. Regarding the EC_{50} calculated by comparing the results with those described in Table 3, both values lie within the range indicated as slightly toxic for soils in this region.

Results of low or null toxicity level as determined by the Microtox bioassay have been previously reported for hydrocarbon contaminated clayed soils from the same region (Adams et al. 1999), that may be due to a strong degree of sorption of diesel into the soil, in addition to some degree of degradation of the residual compounds. In our case, biodegradation and weathering processes occurring in both treatments probably generated compounds that increased the toxicity of soil leachates, as compared to the initial TU=2.0 value, equivalent to an EC_{50} of 179,317 ppm.

http://dx.doi.org/10.5772/56233

Treatment	Toxicity (TU)*	EC_{50} (ppm)
Bioaugmented	13.7 ± 2.8	75 218 ± 17 679
Non-bioaugmented	17.0 ± 0.2	59 010 ± 762.2

*Toxicity unit = $(1/EC_{50})$ x10^6.

Table 3. Toxicity analysis results for the artificially diesel-contaminated soil, under bioaugmented and non-bioaugmented treatments, after 90 days of culture.

3.3. Biodegradation assay of Olmeca crude oil in soil

Among the different types of oil produced in Mexico are the Maya heavy oil with gravity of 22° API (3.3% sulfur by weight), the Istmo light oil with 33.6° API (1.3% of sulfur by weight) and the Olmeca super light oil with API gravity of 39.3 ° (0.8% sulfur by weight). Olmeca and Maya oils are produced in oil fields located in the southeastern region of the country, in the states of Veracruz, Tabasco and Chiapas. There are several sites affected by the accidental discharge of either Maya and Olmeca oils or their refined products. The removal efficiency of the co-culture (CC) of degrading bacteria and fungi was evaluated by addressing residual total petroleum hydrocarbons (TPH) from Olmeca crude oil. The CC was the same used in the diesel biodegradation experiments.

3.3.1. Biodegradation kinetics of Olmeca crude oil in soil

The results corresponding to biodegradation and concentrations of residual and consumed hydrocarbons during the 90 days assay are shown in Figure 3. Among the four treatments tested, both bioaugmented treatments recorded the highest levels of biodegradation (panels a and c), being 58.3 ± 3% and 55.8 ± 2%, respectively. While in the non-bioaugmented treatments (panels b and d) the final biodegradation was about 48 ± 2% and 37.6 ± 2%, respectively.

The period of increased oil consumption corresponded to days 1 to 45 of assay, for the bioaugmented treatments. This consumption was less accelerated for NB-SS, where after 22 days of treatment, oil removal decreased considerably (Figure 3b). Furthermore, in the B-NSS treatment (Figure 3c), there is a marked decrease of the residual hydrocarbon concentration, between days 1 to 60, which is less pronounced in the non bioaugmented treatments (Figure 3b and 3d). These two treatments also showed a marked decrease in the oil removal trend from day 60 to the end of the assay (90 days). In the case of bioaugmented treatments, in this same period (60 to 90 days) was still detected a slight consumption of oil.

Values for residual hydrocarbon content at the end of the assay were statistically analyzed (ANOVA, F-test, $p<0.05$), bioaugmented treatments showed significant differences compared to the non- bioaugmented. However, no differences were found between both bioaugmented treatments that presented similar behavior.

Regarding the number of microorganisms, all treatments remained at levels higher than 1x10^6 CFU g^{-1} dry soil. Maximum values were obtained at 60 days in the bioaugmented trays (ca 1x10^8 CFU g^{-1} dry soil). Fertilizer was added to all trays at the start of assay, this fact probably limited the efficiency of biodegradation, as after 60 days of assay there was a decrease in the trend of oil removal (Fig. 3a – 3c).

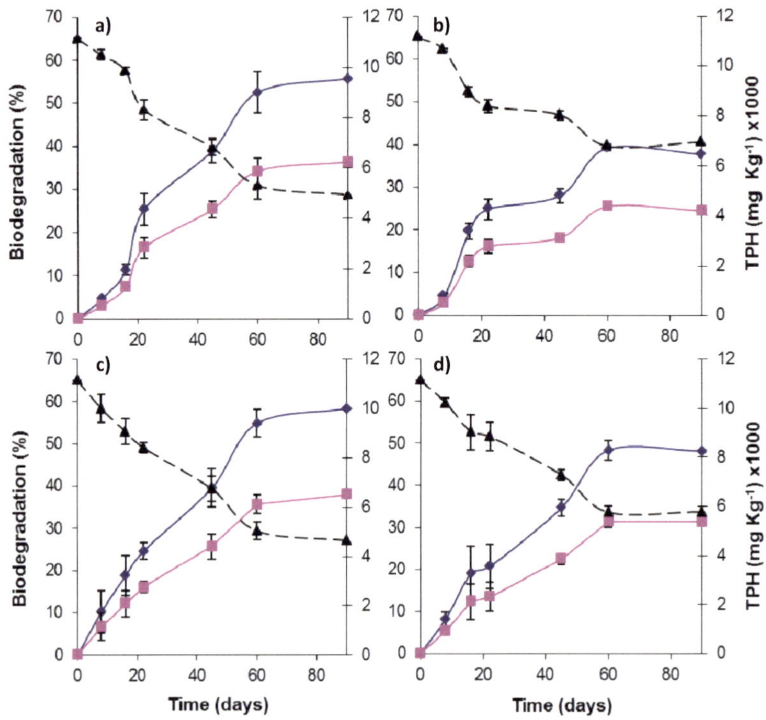

Figure 3. Kinetics of hydrocarbons consumption in soil under different treatments: a) bioaugmented - sterile soil (B-SS), b) non bioaugmented - sterile soil (NB-SS), c) bioaugmented – non sterile soil (B-NSS), d) non bioaugmented – non sterile soil (NB-NSS). Curves represent values of biodegradation (♦), consumed hydrocarbons (■) and residual hydrocarbons (▲). The values are the average of three replicates at each sampling time. The data were corrected considering extractable fat content of the soil.

3.3.2. Kinetic parameters for biodegradation of Olmeca crude oil

With the hydrocarbons consumption data we calculated the following consumption rates: a) global consumption rate; b) maximum consumption rate. Both parameters are volumetric values representing the biodegradation rate allowing a more representative comparison (Diaz-Ramirez et al. 2008). Data are presented in Table 4.

Samples of the B-NSS treatment reached maximum consumption rate after 22 days of experiment (data not shown), while for samples B-SS and NB-SS (sterile soils) the maximum rate was obtained later (between 22 and 45 days).

Treatment	Biodegradation (%)	GCR [a] (mg kg^{-1} d^{-1})	MCR [b] (mg kg^{-1} d^{-1})
B-SS	55.8 ± 2.0	69.1	259.4
NB-SS	37.6 ± 2.3	46.6	210.6
B-NSS	58.3 ± 3.2	72.3	142.6
NB-NSS	48.1 ± 2.3	59.6	150.5

[a] Global Consumption Rate (GCR) calculated as:

$$GCR = \frac{\left[TPH_{final}\right] - [TPH_{inital}]}{t_0 - t_{final}} \tag{2}$$

Where:

GCR: Global consumption rate

TPH_{final}= Total petroleum hydrocarbon concentration at the end of the assay

TPH_{inital}= Initial total petroleum hydrocarbon concentration

t_0= initial time

t_{final}= final experimental time

[b] Maximum Consumption Rate (MCR) calculated as:

$$MCR = \frac{\left[TPH_r\right]_n - [TPH_r]_{n-1}}{t_{n-1} - t_n} \tag{3}$$

Where:

MCR: Maximum consumption rate determined between each sampling interval

$[TPH_r]_n$= Residual total petroleum hydrocarbon concentration at t_n

$[TPH_r]_{n-1}$= = Residual total petroleum hydrocarbon concentration at t_{n-1}

t_n= Sampling at "t" time

t_{n-1}= Sampling time previous to t_n

Table 4. Olmeca oil biodegradation, global and maximum consumption rates, calculated for the different soil treatments after 90 days of assay.

In order to calculate quantitative kinetic parameters, the obtained quantitative data were adjusted to a first order model equation (see section 2.2.2 Biodegradation kinetics of diesel in soil). Table 5 presents the biodegradation constant and half life time obtained from the first-order linear fit performed for the different treatments.

The better fit to first order model was achieved with the data obtained from bioaugmented treatments (B-SS, B-NSS) and even for NB-NSS (r^2> 0.94), in contrast, residual hydrocarbon concentration in NB-SS samples showed a non-linear behavior (r^2 ~ 0.7).

Similarly to the biodegradation experiments of diesel in soil, it was determined a degradation constant (-k) and the half life time average data of three replicate trays are presented in Table 5.

Treatment	-k (d^{-1})	Half life time (d)
B-SS	-0.0103	67.5 ± 5.0
NB-SS	-0.0067	103.6 ± 5.2
B-NSS	-0.0110	63.0 ± 1.0
NB-NSS	-0.0088	79.1 ± 1.4

Table 5. Biodegradation constant and half-life time for Olmeca crude oil in soil, after a 90-day assay.

The biodegradation constant values were consistent with the results described, being higher (1.2 to 1.5 times) in the bioaugmented treatments (B-SS, B-NSS). The half life time was over 100 days (3.4 months) to the NB-NSS, while this value was slightly above 60 days (2 months) in the bioaugmented treatments. This indicates that it would take about 4 months to achieve complete biodegradation of the carbon source by applying the tested inoculum, regardless the soil is previously sterilized. For TPH (1,010 – 2,141 mg kg^{-1}), first-order biodegradation rate constants were obtained by Chang et al. (2010) in similar range (0.011 - 0.024 d^{-1}).

It is noteworthy that as in the case of diesel biodegradation (Figure 1), with Olmeca crude oil also was observed a decrease in the consumption rate after 60 days of assay (Figure 3). It is likely due to the reduction of the easily assimilable compounds after this period, staying in the soil compounds more resistant to biodegradation. Another possibility is a reduction in the availability of secondary nutrients and minerals, provided by the sugar cane bagasse and commercial fertilizer, respectively, restricting the removal of oil compounds still present.

The results obtained are comparable to those obtained in similar studies that used Diesel as carbon source and added inocula to promote biodegradation. Margesin et al, (2007) reported similar results, finding that with concentrations of 10,000 mg TPH (diesel) Kg^{-1} soil, the biodegradation was 31% to 53% in 38 days, assayed soil was biostimulated using inorganic fertilization with NPK nutrients (C/N ratio of 20:1) or oleophilic fertilization with Inipol EAP22. Mishra et al, (2001), tested the inorganic nutrient addition and the bioaugmentation using a defined mixed culture composed by five oily-sludge-degrading indigenous bacterial strains, these strains belong to genus *Acinetobacter (two strains), Burkholderia and Pseudomonas*, one of them was not identified. It inoculum was used for *in situ* bioremediation of hydrocarbons, finding a 48.5% TPH decrease in presence of the defined mixed culture and 17% in the biostimulated (non-inoculated) treatment after 120 days.

3.4. Monitoring of Natural Attenuation (MNA) of contaminated soil and sediments in a riparian environment

A severe gasoil spill (TPH-medium fraction) affected soils and rivers in a mountainous area southeast Mexico. In this tropical region most soils have silt-clayed texture characteristics, being the main human activities livestock and rain fed agriculture. Another distinctive feature of this area is its diverse basins containing most of the country's water resources. The oil spill occurred on a site adjacent to a river system consisting of two streams and a river that flows into the upper Coatzacoalcos River basin. The first of the streams was the most affected due to the increased flow of hydrocarbons. The stream surrounding area has certain degree of conservation and ecological characteristics of a healthy forest area, with silty riverbed and sediments. Based on current Mexican regulations for soils contaminated with hydrocarbons, characterization and site diagnosis was required to propose alternatives for restoration of the affected areas. Particular interest was put in determining gasoil passive biodegradation potential under field conditions, in a series of points located along the stream and close to the spill site. Table 6 shows the results of three points selected for MNA of the gasoil spill occurred in this area.

Sampling point	Sampling time (d)	Gasoil concentration (mg Kg^{-1})[a]	Biodegradation extent (%)[b]	Half life time (d)[c]
P 2	0	59 626	0	
P 2	36	44 092	26.0	
P 2	87	9 670	83.8	
P 2	117	ND	ND	
P 2	383	1 985	96.6	73.0
P 9	0	9 627		
P 9	36	12 682		
P 9	87	3 846	60.5	
P 9	117	3 583	62.8	
P 9	383	3 351	65.2	192.5
P 10	0	34 678	0	
P 10	36	14 817	57.2	
P 10	87	12 033	65.3	
P 10	117	7 486	78.4	
P 10	383	6 715	80.6	51.7

[a] Residual gasoil determined as medium fraction (C10 – C28) EPA 8015b.

[b] Biodegradation calculated considering the initial content of gasoil.

[c] Half life time obtained from first order adjustment of the soil residual concentration data.

Table 6. MNA parameters of gasoil measured at three sampling points of the spill-impacted area.

Sediment samples and superficial soil (0-10 cm) were collected from sampling points 9 and 10 (P9 and P10). Soil samples from 30 - 45 cm depth were taken from sampling point 2 (P2). Residual hydrocarbons data were well described by the first order model having a high lineal correlation ($r^2 = 0.926$ to 0.937). Additionally, in P2 hydrocarbons were detected in the stream-bed sediments (17 100 mg kg^{-1}), with a later 84% reduction (2 662 mg kg^{-1}) in 81 days. For some of the samples we found heterogeneous results, even finding increases from baseline (P9, times 0 to 36 d), this is explained by considering the inherent field variability on the presence of hydrocarbons and the stream dynamics in the site. However, the general behavior of the residual hydrocarbons concentration was towards an accelerated biodegradation, with reductions between 60 to 80% compared to the initial concentrations. According to the half-life calculated from gasoil biodegradation data, the residual concentrations will reach levels below the permissible limit established in current Mexican regulations (<1 200 mg kg^{-1}) after 6 – 9 months.

The calculated global consumption rates (GCR) ranged between 51 mg kg^{-1} soil d^{-1} for a low contaminated site to 232 mg kg^{-1} soil d^{-1}, for the most contaminated (P2). While average maximum consumption rates for P10 and P2 scored 551 and 675 mg kg^{-1} d^{-1}, respectively, these values were achieved between 36 and 87 days of monitoring. Regarding the presence of PAHs, in most of the studied sampling points the levels were below the quantification threshold for CG analysis, except for benzo [a] pyrene that reached 0.884 - 2.76 mg kg^{-1}.

In general, the numbers of diesel degraders and heterotrophic bacteria and fungi were high in the sampled soils. Values for degraders ranged from 10^4 to 10^7, corresponding to 0.1 to 30% of the total microbial population. This range of degraders is even larger than previous reports. Prince, (2003) found that the number of oil-degrading bacteria in a contaminated soil was around 0.1 to 10% of the total microbial population. Our results can be attributed to the favorable environmental conditions detected on the walls of the stream, with abundant vegetation, high soil moisture, rich in organic matter that has been previously deposited by water flow.

Regarding sediment samples, moderate numbers of oil-degrading microorganisms (10^3 to 10^6) were found, however, a high removal of hydrocarbons was scored. This may be explained by desorption of the gasoil present in sediments, due to washing effect of water flow. Subsequently, biodegradation could take place in the liquid phase.

Accelerated biodegradation rates were observed for most of the monitored points resulting in significant reduction of gasoil concentration after two to three months (short half-life), even at the points of highest content. On view of these results, monitored natural attenuation can be considered as an option for site restoration. Although a site characterization strategy is needed, in order to determine environmental conditions that may influence hydrocarbons disappearance, in addition to the sole presence of an oil-degrading community. MNA strategies could be applied to other sites of comparable ecological value or similar tropical regions.

4. Concluding remarks

Diesel biodegradation in liquid media was higher in presence of the co-culture (bacterial and fungal standardized mixture) and the F-DMC. These previously characterized microorgan-

http://dx.doi.org/10.5772/56233

isms showed high capacity for diesel biodegradation in liquid medium and in soil (120,000 mg kg^{-1} soil). Limitations in biodegradation extent probably were due to depletion of nutrients necessary for the growth or activity of the oil-degrading microorganism. Another possible cause was the low bioavailability of diesel (soil weathering), which affected the activity of both native and exogenous degrading microorganisms. Based on the conspicuous colony morphology of the introduced bacterial strains *B. subtilis* 7A and *G. rubripertincta* during biodegradation in soil assays, they prevailed at the end of the experiments, showing a good degrading capacity and ability to compete with native soil bacteria.

The high diesel and Olmeca crude oil degrading activities registered in both liquid medium and soil assays are promising alternatives for the application of this type of co-culture as a part of active remediation strategies for contaminated soils based, together with biostimulation and bioaugmentation approaches.

Finally, MNA schemes like the one here assayed, could be useful for contaminated sites with a good ecological conservation degree, where an active remediation technique may result in more damage and increased costs. Special attention should be put in the physicochemical effects of hydrocarbons on the soil properties (i e. field capacity, water repellency, salinization, lixiviate residual toxicity). These approaches may favour re-establishment of vegetation and fertility of the soil matrix.

Acknowledgements

Financial support is acknowledged to the Mexican Ministry for Public Education (Project PROMEP No. 20110123) and Juárez Autonomous University of Tabasco (Project PFICA No. 2012010). Special thanks are given to Cornelio Contreras Juárez, Francisco Guzmán Osorio and Jorge Ortiz Maya for experimental and field assistance.

Author details

Ildefonso Díaz-Ramírez[1], Erika Escalante-Espinosa[1], Randy Adams Schroeder [1], Reyna Fócil-Monterrubio[1] and Hugo Ramírez-Saad[2*]

*Address all correspondence to: hurasa@correo.xoc.uam.mx

1 Academic Division of Biological Sciences, Juarez Autonomous University of Tabasco, Tabasco, Mexico

2 Departament of Biological Systems, Metropolitan Autonomous University – Xochimilco, Mexico City, Mexico

References

[1] Adams RH, Castillo-Acosta O, Escalante-Espinosa E, Zavala-Cruz J. 2011. Natural attenuation and phytoremediation of petroleum hydrocarbon impacted soil in tropical wetland environments, In: Torres LG, Bandala ER. (Eds). Remediation of Soils and Aquifers. pp 1-24. Nova Publishers, New York.

[2] Adams RH, Domínguez-Rodríguez VI, García Hernández L. 1999. Potencial de la biorremediación de suelo y agua impactado por petróleo en el trópico Mexicano. Terra 17(2): 159-174.

[3] Adams RH, Guzmán-Osorio FJ. 2008. Evaluation of land farming and chemical-biological stabilization for treatment of heavily contaminated sediments in a tropical environment. International Journal Environmental Science Technology 5(2): 169-178.

[4] Adams RH, Zavala Cruz J, Morales García FA. 2008. Concentración residual de hidrocarburos en suelo del trópico. II: Afectación a la fertilidad y su recuperación. Interciencia. 33: 483 -489.

[5] Amann R, Kühl M. 1998. *In situ* methods for assessment of microorganism and their activities. Current Opinion in Microbiology 1: 352-358.

[6] Boettcher G, Nyer EK. 2001. *In situ* Bioremediation. In: Nyer EK, Palmer PL, Carman EP, Boettcher G, Bedessman JM, Lenzo F, Crossman T L, Rorech GJ, Kidd DF (Eds). *In situ* Treatment Technology. pp. 261-281, Arcadis, Geraghty & Miller. FL. USA.

[7] Boonchan S, Britz ML, Stanley GA. 2000. Degradation and mineralization of high-weight polycyclic aromatic hydrocarbons by defined fungal-bacterial cocultures. Applied and Environmental Microbiology 66(3): 1007-1019.

[8] Cardona S, Iturbe R. 2003. Degradación de diésel Mexicano por un consorcio de bacterias de un suelo agrícola. Revista Dyna 70(138):13-26. Available at: http://redalyc.uaemex.mx/pdf/496/49613802.pdf

[9] Chang W, Dyen M, Spagnuolo L, Simon P, Whyte L, Ghoshal S. 2010. Biodegradation of semi- and non-volatile petroleum hydrocarbons in aged, contaminated soils from a sub-Arctic site: Laboratory pilot-scale experiments at site temperatures. Chemosphere 80: 319–326.

[10] Chemlal R, Tassist A, Drouiche M, Lounici H, Drouiche N, Mameri N. 2012. Microbiological aspects study of bioremediation of diesel-contaminated soils by biopile technique. International Biodeterioration and Biodegradation 75: 201-206.

[11] Díaz-Ramírez IJ, Escalante-Espinosa E, Favela-Torres E, Gutiérrez-Rojas M, Ramírez-Saad H. 2008. Design of bacterial defined mixed cultures for biodegradation of specific crude oil fractions, using population dynamics analysis by DGGE. International Biodeterioration and Biodegradation 62: 21-30.

[12] Díaz-Ramírez IJ, Gutiérrez-Rojas M, Ramírez-Saad H, Favela-Torres E. 2003. Biodegradation of Maya crude oil fractions by bacterial strains and a defined mixed culture isolated from *Cyperus laxus* rhizosphere soil in a contaminated site. Canadian Journal of Microbiology 49: 755-761.

[13] Díaz-Ramírez IJ. 2000. Biodegradación de hidrocarburos por un consorcio microbiano de la rizósfera de una planta nativa de pantano. Tesis de Maestría, Universidad Autónoma Metropolitana – Iztapalapa, México.

[14] EPA (USA) 8015, 1996. Non-halogenated organics using GC / FID. SW 846. Test methods for evaluating soil waste, physical/chemical methods. Available at: http://www.caslab.com/EPA-Methods/PDF/8015b.pdf

[15] EPA (USA) 1999. Use of Monitored Natural Attenuation at Superfund, RCRA Corrective Action, and Underground Storage Tank Sites. OSWER Directive number: 9200.4-17P. Available at: http://www.epa.gov/oust/directiv/d9200417.pdf

[16] Escalante-Espinoza E. 2005. Estudio de la capacidad fitorremediadora de *Cyperus laxus* Lam. en suelo contaminado con hidrocarburos. Tesis de Doctorado, Universidad Autónoma Metropolitana – Iztapalapa, México.

[17] Eweis JB, Ergas SJ, Chang, DP. Schroeder ED. 1998. Bioremediation Principles. pp. 88 - 107. WCB/ McGraw Hill International.

[18] Greenwood PF, Wibrow S, George SJ, Tibbett M. 2009. Hydrocarbon biodegradation and soil microbial community response to repeated oil exposure. Organic Geochemistry. 40(3): 293-300.

[19] Hye HJ, Kim JK, Choi O, Kyung-Suk C, Hee WR. 2005. Characterization of a diesel-degrading bacterium, *Pseudomonas aeruginosa* IU5, isolated from oil-contaminated soil in Korea. World Journal of Microbiology y Biotechnology 21:381-384.

[20] Ko SH. Lebeault JM. 1999. Effect of a mixed culture on co-oxidation during the degradation of saturated hydrocarbon mixture. Journal of Applied Microbiology 87: 72-79.

[21] Leahy JG, Colwell RR. 1990. Microbial degradation of hydrocarbons in the environment. Microbiological Reviews 54: 305-315.

[22] Liu GPW, Chang TC, Whang LM, Kao CH, Pan PT, Cheng SS. 2011. Bioremediation of petroleum hydrocarbon contaminated soil: Effects of strategies and microbial community shift. International Biodeterioration and Biodegradation 65(8): 1119-1127.

[23] Liu S, Suflita JM. 1993. Ecology and evolution of microbial populations for bioremediation. Trends in Biotechnology 11: 344-352.

[24] Lorch HJ, Benckieser G. Ottow JCG. 1995. Basic methods for counting microorganisms in soil and water. In: Alef K, Nannipieri P (Eds). Methods in Applied Soil Microbiology and Biochemistry, pp. 146-161, Academic Press, Great Britain.

[25] Madsen EL. 1998. Epistemology of environmental microbiology. Environmental Science Technology.32(4): 429-439.

[26] Madueño L, Coppotelli BM, Alvarez HM, Morelli IS. 2011. Isolation and characterization of indigenous soil bacteria for bioaugmentation of PAH contaminated soil of semiarid Patagonia, Argentina. International Biodeterioration and Biodegradation 65: 345-351.

[27] Margesin R, Hämmerle M, Tscherko D. 2007. Microbial activity and community composition during bioremediation of diesel-oil-contaminated soil: effects of hydrocarbon concentration, fertilizers, and incubation time. Microbial Ecology 53: 259–269.

[28] Margesin R. 2005. Determination of Enzyme Activities in Contaminated Soil. In: Margesin R, Schinner F (Eds). Manual for Soil Analysis – Monitoring and Assessing Soil Bioremediation. Soil Biology, Vol. 5. pp. 309-319. Springer, Germany.

[29] Marquez-Rocha FJ, Hernández RV, Lamela MT. 2001. Biodegradation of diesel oil in soil by a microbial consortium. Water, Air and Soil Pollution 128: 313–320.

[30] Medina-Moreno SA, Huerta-Ochoa S, Gutiérrez-Rojas M. 2005. Hydrocarbon biodegradation in oxygen limited sequential batch reactors by consortium from weathered oil-contaminated soil. Canadian Journal of Microbiology 51: 231-239.

[31] Mishra S, Jyot J, Kuhad RC, Lal B. 2001. Evaluation of inoculum addition to stimulate *in situ* bioremediation of oily-sludge-contaminated soil. Applied and Environmental Microbiology 67: 1675–1681.

[32] Mohammed D, Ramsubhag A, Beckles DM. 2007. An assessment of the biodegradation of petroleum hydrocarbons in contaminated soil using non-indigenous, commercial microbes. Water Air and Soil Pollution 182: 349–356.

[33] NMX-AA 112-SECOFI, 1995. Norma Oficial Mexicana. Water and sediment analysis – Acute Toxicity Evaluation with *Photobacterium phsophoreum* – Test method. Available at: http://www.imta.gob.mx/cotennser/images/docs/NOM/NMX-AA-112-1995.pdf

[34] NOM-021-SEMARNAT-2000. Norma Oficial Mexicana, que establece las especificaciones de fertilidad, salinidad, y clasificación de suelos, estudio, muestreo y análisis. México, D., F., Diario Oficial de la Federación. Martes 23 de Abril del 2003.

[35] NOM-053-ECOL-1993. Norma Oficial Mexicana. Procedimiento para llevar a cabo la prueba de extracción para determinar los constituyentes que hacen a un residuo peligroso por su toxicidad al ambiente. Available at: http://biblioteca.semarnat.gob.mx/janium/Documentos/Ciga/agenda/PPD02/DO2283m.pdf

[36] NOM-138-SEMARNAT/SS-2003. Norma Oficial Mexicana. Límites máximos permisibles de hidrocarburos en suelos y las especificaciones para su caracterización y remediación. Diario Oficial de la Federación, México, D.F. 29-Mar-2005.

[37] Petróleos Mexicanos (PEMEX). 2008. Subdirección de auditoría en seguridad industrial y protección ambiental. Hoja de datos de seguridad de sustancias (Diésel UBA).

[38] Prince RC. 2003. Biodegradation of petroleum and other hydrocarbons. In: Encyclopedia of Environmental Microbiology. John Wiley & Sons, Inc. New Jersey. Available at (http://www.mrw.interscience.wiley.com/eem/articles/env118).

[39] Richard JY, Vogel TM. 1999. Characterization of a soil bacterial consortium capable of degrading diesel fuel. International Biodeterioration and Biodegradation 44: 93-00.

[40] Roldán-Carrillo T, Castorena-Cortés G, Zapata-Peñasco I, Reyes-Avila J, Olguín-Lora P. 2012. Aerobic biodegradation of sludge with high hydrocarbon content generated by a Mexican natural gas processing facility. Journal of Environmental Management 95: S93-S98.

[41] Sabaté J, Viñas M. Solanas AM. 2004. Laboratory-scale bioremediation experiments on hydrocarbon-contaminated soils. International Biodeterioration and Biodegradation 54: 19-25.

[42] Sayara T, Borràs E, Caminal G, Sarrà M, Sánchez A, 2011. Bioremediation of PAHs-contaminated soil through composting: Influence of bioaugmentation and biostimulation on contaminant biodegradation. International Biodeterioration and Biodegradation 65: 859-865.

[43] Thouand G, Bauda P, Oudot J, Kirsch G, Sutton C, Vidalie JF. 1999. Laboratory evaluation of crude oil biodegradation with commercial or natural microbial inocula. Canadian Journal of Microbiology 45: 106-115.

[44] Torsvik V, Ovreas L. 2002. Microbial diversity and function in soil: from genes to ecosystems. Current Opinion in Microbiology 5: 240-245.

[45] Vallejo V, Salgado L, Roldán F. 2005. Evaluación de la estimulación en la biodegradación de HTPs en suelos contaminados con petróleo. Revista Colombiana de Biotecnología 3(2): 67-68.

[46] Van der Meer JR, De Vos WM, Harayama S, Zehnder AJB. 1992. Molecular mechanisms of genetic adaptation to xenobiotic compounds. Microbiological Reviews 56 (4): 677-694.

[47] Van Hamme JD, Odumeru JA, Ward OP. 2000. Community dynamics of a mixed-bacterial culture growing on petroleum hydrocarbons in batch culture. Canadian Journal of Microbiology 46: 441-450.

[48] Van Hamme JD, Singh A, Ward OP. 2003. Recent advances in petroleum microbiology. Microbiology and Molecular Biology Reviews 67(4): 503-549.

[49] Viñas M, Grifoll M, Sabaté J, Solanas AM. 2002. Biodegradation of a crude oil by three microbial consortia of different origins and metabolic capabilities. Journal of Industrial Microbiology and Biotechnology 28: 252–260.

[50] Watanabe K, Hamamura N. 2003. Molecular and physiological approaches to understanding the ecology of pollutant degradation. Current Opinion in Microbiology 14: 289-295.

[51] Wrenn BA, Venosa AD. 1996. Selective enumeration of aromatic and aliphatic hydrocarbon degrading bacteria by a most-probable-number procedure. Canadian Journal of Microbiology 42: 252-258.

[52] Zhang D, Mörtelmaier C, Margesin R. 2012. Characterization of the bacterial archaeal diversity in hydrocarbon-contaminated soil. Science of the Total Environment 421: 184–196.